CAMBRIDGE LIBRARY COLLECTION

Books of enduring scholarly value

Astronomy

From ancient times, humans have tried to understand the workings of the world around them. The roots of modern physical science go back to the very earliest mechanical devices such as levers and rollers, the mixing of paints and dyes, and the importance of the heavenly bodies in early religious observance and navigation. The physical sciences as we know them today began to emerge as independent academic subjects during the early modern period, in the work of Newton and other 'natural philosophers', and numerous sub-disciplines developed during the centuries that followed. This part of the Cambridge Library Collection is devoted to landmark publications in this area which will be of interest to historians of science concerned with individual scientists, particular discoveries, and advances in scientific method, or with the establishment and development of scientific institutions around the world.

An Original Theory or New Hypothesis of the Universe, Founded upon the Laws of Nature

Although his yeoman father is said to have burnt his books to discourage excessive studiousness, Thomas Wright (1711–86) nevertheless acquired considerable knowledge in the fields of mathematics, navigation and astronomy. Later benefiting from the patronage of wealthy families, he also surveyed estates, designed gardens, and tutored aristocrats. He is best known, however, for his contribution to astronomy: this illustrated work of 1750 was his most famous publication. Written in the form of nine letters, the book quotes both poets and scientists in the opening discussion as Wright sets out to fuse, rather than separate, science and religion. Combining his observations of the Milky Way with his theological belief in a universe of perfect order, he notes, among other things, that our galaxy appears to be disc-shaped. While largely ignored by contemporary astronomers, Wright's ideas can be seen as a forerunner to more sophisticated conceptions of our galaxy's configuration.

Cambridge University Press has long been a pioneer in the reissuing of out-of-print titles from its own backlist, producing digital reprints of books that are still sought after by scholars and students but could not be reprinted economically using traditional technology. The Cambridge Library Collection extends this activity to a wider range of books which are still of importance to researchers and professionals, either for the source material they contain, or as landmarks in the history of their academic discipline.

Drawing from the world-renowned collections in the Cambridge University Library and other partner libraries, and guided by the advice of experts in each subject area, Cambridge University Press is using state-of-the-art scanning machines in its own Printing House to capture the content of each book selected for inclusion. The files are processed to give a consistently clear, crisp image, and the books finished to the high quality standard for which the Press is recognised around the world. The latest print-on-demand technology ensures that the books will remain available indefinitely, and that orders for single or multiple copies can quickly be supplied.

The Cambridge Library Collection brings back to life books of enduring scholarly value (including out-of-copyright works originally issued by other publishers) across a wide range of disciplines in the humanities and social sciences and in science and technology.

An Original Theory or New Hypothesis of the Universe, Founded upon the Laws of Nature

*And Solving by Mathematical Principles
the General Phænomena of the Visible Creation,
and Particularly the Via Lactea*

Thomas Wright

CAMBRIDGE
UNIVERSITY PRESS

CAMBRIDGE
UNIVERSITY PRESS

University Printing House, Cambridge, CB2 8BS, United Kingdom

Cambridge University Press is part of the University of Cambridge.

It furthers the University's mission by disseminating knowledge in the pursuit of
education, learning and research at the highest international levels of excellence.

www.cambridge.org
Information on this title: www.cambridge.org/9781108073745

© in this compilation Cambridge University Press 2014

This edition first published 1750
This digitally printed version 2014

ISBN 978-1-108-07374-5 Paperback

A N
ORIGINAL THEORY
O R
NEW HYPOTHESIS
OF THE
UNIVERSE,

Founded upon the

LAWS of NATURE,

AND SOLVING BY

MATHEMATICAL PRINCIPLES
T H E

General PHÆNOMENA of the VISIBLE CREATION;

AND PARTICULARLY

The VIA LACTEA.

Compris'd in Nine Familiar LETTERS from the AUTHOR to his FRIEND.
And Illuſtrated with upwards of Thirty Graven and Mezzotinto Plates,
By the Beſt MASTERS.

By THOMAS WRIGHT, of DURHAM.

One Sun by Day, by Night ten Thouſand ſhine,
And light us deep into the DEITY.　　Dr. YOUNG.

LONDON:
Printed for the AUTHOR, and ſold by H. CHAPELLE, in *Groſvenor-Street.*
MDCCL.

THE

PREFACE

T HE Author of the following Letters having been flattered into a Belief, that they may probably prove of some Use, or at least Amusement to the World, he has ventured to give them, at the Request of his Friends, to the Publick. His chief Design will be found an Attempt towards solving the Phænomena of the *Via Lactea*, and in consequence of that Solution, the framing of a regular and rational Theory of the known Universe, before unattempted by any. But he is very sensible how difficult a Task it is to advance any new Doctrine with Success, those who have hitherto attempted to propagate astronomical Discoveries in all Ages, have been but ill rewarded for their Labours, tho' finally they have proved of the greatest Benefit and Advantage to Mankind. This ungrateful Lesson we learn from the Fate of those ingenious Men, who, in ignorant Times, have unjustly suffered for their superior Knowledge and Discoveries; they who first conceived the Earth a Ball, were treated only with Contempt for their idle and ridiculous Supposition, as it was called; and he who first attempted to explain the *Antipodes*, lost his Life by it; but in this Age Philosophers have nothing to fear of this sort, the great Disadvantages attending Authors now, are of a widely different

A 2   Nature,

Nature, rifing from the infinite Number of Pretenders to Knowledge in this Science, and much is to be apprehended from improper Judges, tho' from real ones nothing; for nothing is more certain than this, as much as any Subject exceeds the common Capacity of Readers, fo much will the Work in general be condemned; the Air of Knowledge is at leaft in finding Fault, and this vain Pretence generally leads People, who have no real Foundation for their Judgment to argue from, to ridicule what they are too fenfible they do not underftand.   Thus the fame Difadvantages too often attend both in publick and private an exceeding good Production equally the fame as a very bad one : But the Author is not vain enough to think this Work without Faults, has rather Reafon to fear, from the Weaknefs of his own Capacity, that there may be many; but he hopes the Defign of the Whole will, in fome meafure, plead for the Imperfection of the Parts, if the Merits of the Plan fhould be found infufficient for his full Pardon, in attempting fo extenfive a Subject.

In a Syftem thus naturally tending to propagate the Principles of Virtue, and vindicate the Laws of Providence, we may indeed fay too little, but cannot furely fay too much; and to make any further Apology for a Work of fuch Nature, where the Glory of the Divine Being of courfe muft be the principal Object in View, would be too like rendering Virtue accountable to Vice for any Author to expect to benefit by fuch Excufe.   The Motive which induces us to the Attempt of any Performance, where no good Reafon can be fuppofed to be given for the Omiffion, or Neglect of it, will always be judged an unneceffary Promulgation, and confequently every Attempt towards the Difcovery of Truth, the Enlargement of our Minds, and the Improvement of our Underftandings will naturally become a Duty.   If therefore this Undertaking falls fhort of being inftrumental towards the advancing the Adoration of the Divine Being in his infinite Creation of higher Works, and proves unable to anfwer all Objections that may poffibly arife againft it, yet will its Imperfections appear of fuch a Nature to every candid Reader, as to afford the Author a fufficient Apology for producing them to the World : And it is to be hoped farther, that where a Work is entirely upon a new Plan, and the Beginning, as it were, of a new Science, before unattempted in any Language, the Author having dug all his Ideas from the Mines of Nature, is furely intitled to every kind of Indulgence.

To

To thofe who are weak enough to think that fuch Enquiries as thefe are over-curious, vain, and prefumptive, and would willingly, fuitable to their own Ignorance and Comprehenfion, fet Bounds to other People's Labours, I anfwer with Mr. *Huygens*, " That if our Forefathers had " been at this Rate fcrupulous, we might have been ignorant ftill of the " Magnitude and Figure of the Earth ; or that there was fuch a Place as " *America*. We fhould not have known that the Moon is enlightened by " the Sun's Rays, nor what the Caufes of the Eclipfes of each of them " are ; nor a Multitude of other Things brought to Light by the late " Difcoveries in Aftronomy ; for what can a Man imagine more abftrufe, " or lefs likely to be known, than what is now as clear as the Sun."

> Had we ftill paid that Homage to a Name,
> Which only God and Nature juftly claim ;
> The weftern Seas had been our utmoft Bound,
> Where Poets ftill might dream the Sun was drown'd;
> And all the Stars that fhine in Southern Skies,
> Had been admir'd by none but favage Eyes.
>
> DRYDEN.

Befides the Noblenefs and Pleafure of thefe Studies, *Wifdom* and *Morality* are naturally advanced, and much benefited by them, and even Religion itfelf receives a double Luftre, " to the Confufion of thofe who " would have the Earth, and all Things formed by the fhuffling Concourfe " of Atoms, or to be without Beginning." In Aftronomy, as well as in natural Philofophy, though we cannot pofitively affirm every thing we fay to be Facts and Truth, yet in fo noble and fublime a Study as that of *Nature*, it is glorious, as Mr. *Huygens* fays, even to arrive at Probability.

Notwithftanding then the Difadvantages which ever have attended all new Difcoveries, either thro' the Ignorance of the Age, or the univerfal Paffion of Ridicule in fuch contented Creatures, as can't comprehend, yet ever attacking with a fool-hardy Refolution, the advancing Enfigns of Knowledge, if Ignorance was Virtue, and Wifdom Vice ; I fay, regardlefs of this noify Shore, it is fure our Duty to fpring forward, and explore the fecret Depths of Infinity, and the wonderful hidden Truths of this vaft Ocean of Beings. But how the heavenly Bodies were made, when they were

made

made, and what they are made of, and many other Things relating to their Entity, Nature, and Utility, feems in our prefent State not to be within the Reach of human Philofophy; but then that they do exift, have final Caufes, and were ordained for fome wife End, is evident beyond a Doubt, and in this Light moft worthy of our Contemplation.

> He who thro' vaft Immenfity can pierce,
> See Worlds on Worlds compofe one Univerfe,
> Obferve how Syftem into Syftem runs,
> What other Planets, and what other Suns;
> What varied Being peoples ev'ry Star;
> May tell why Heav'n made all Things as they are.
>
> POPE.

To expect that fo new an Hypothefis fhould meet with univerfal Approbation, would be an unpardonable Vanity; nor is it reafonable every Reader fhould think the Author obliged to remove all his Prejudices and Partialities, fo far as to give him the perfect Picture of the Univerfe he likes beft. In many Cafes it would be fo far from being better for the World, if all Men judged and thought alike, that Providence feems rather to have guarded againft it as an Evil, than any how to have promoted it as a general Good: But the following Theory regards the Whole rather than Individuals: And the many worthy Authors cited in the Work, who have all greatly favoured this extenfive Way of Thinking, will, I hope, be a fufficient Excufe for forming thefe obvious Conjectures into a Theory, efpecially where fo great a Problem is attempted as the Solution of the *Via Lacteal* Phænomenon, which has hitherto been looked upon as an infurmountable Difficulty. How the Author has fucceeded in this Point, is a Queftion of no great Confequence; he has certainly done his beft; another, no Doubt, will do better, and a third perhaps, by fome more rational Hypothefis, may perfect this Theory, and reduce the Whole to infallible Demonftration: The firft Syftem of the folar Planets was far from a true one, but it led the Way to Perfection, and the laft we can never too much admire. It is well known, that the firft Syftem of the Planets was alfo but a Conjecture, yet none will deny that it was an happy one.

The

The Difcovery of the Magnet Poles; the Government of the Tides; proportional Diftance and Periods of the Planets, &c. have all their Ufes, and undoubtedly were defigned to be known. Ignorance is the Difgrace of Mankind, and finks human Nature almoft to that of Reptiles. Knowledge is its Glory and the diftinguifhing Characteriftic of rational Creatures.

To Enquiries of this fort, then fure we may fay with *Milton*, That

GOD'S OWN EAR LISTENS DELIGHTED.

The Subject is, no Doubt, the nobleft in Nature, and as fuch, will always merit the Attention of the thinking Part of Mankind. Men of Learning and Science, in all Ages, have ever made it their peculiar Study. Towards the latter End of the Republic, and afterwards in the more peaceable Times of *Trajan* and the *Plinys*, we have no Reafon to doubt but that Aftronomy was in the higheft Reputation: And notwithftanding *Greece* had been the chief Seat of the Philofophers, yet may we fuppofe *Rome* in thofe Days little inferior in the Knowledge of the Stars, when we find Men * of the firft Figure in Life become Authors upon the Subject.

We have many Inftances to fhew, that Aftronomy was in the greateft Repute amongft the Antients of all Ranks, and almoft every where looked upon as one of the greateft, if not as one of the firft Qualifications of their beft Men. As a Confirmation of which, we find in the hiftorical Accounts of the *Argives*, a very warm Conteft betwixt the two Sons of *Pelops* 1205 Years before *Chrift*, thus teftified by *Lucian*: When the *Argives*, by publick Confent, had decreed that the Kingdom fhould fall to him of the two, who fhould manifeft himfelf the moft learned in the Knowledge of the Stars, *Thyeftes* thereupon is faid to have made known to them, the Conftellation, or Sign of the *Zodiack* call'd *Aries*: But *Atreus* at the fame time difcovering to them the Courfe of the Sun, with his various Rifing and Setting, demonftrating his Motion to be * contrary to that of the Heavens, or diurnal Motion of the Stars, was thereupon elected King.

* *Cicero* tranflated the Phænomena of *Aratus* into *Latin* Verfe. *Julius Cæfar*, as *Pliny* relates, wrote of Aftronomy in *Greek*, and is faid to have left feveral Books of the Motion of the Stars behind him, derived from the Doctrine of the *Egyptians*. *Ant. Chrif.* 45. He with *Sofigenes* reformed the *Roman* Year, which was firft invented by *Numa Pompilius*. *Germanicus Cæfar* alfo tranflated *Aratus's* Phænomena into *Latin* Verfe *Anno Dom.* 15. *Tiberius* and *Hadrian* are alfo faid to have wrote on Aftronomy.

* Hence arofe the Fable of the Sun's going backwards in the Days of *Atreus*, as if ftruck with Abhorrence of his bloody Banquet. *Vide Ovid's* Metamorphofis.

To

To recite more of the moſt eminent Patrons and Profeſſors of this kind of Learning here, will carry me too far from my preſent Purpoſe; for farther Information therefore, I ſhall refer the inquiſitive Reader, to that curious Catalogue in *Sherburn*'s Sphere of *Manilius,* where ſo many ruling † Men of all Ages and Nations ſwell, and illuſtrate the Number.

In a Word, when we look upon the Univerſe as a vaſt Infinity of Worlds, acted upon by an eternal Agent, and crouded full of Beings, all tending through their various States to a final Perfection, and reflect upon the many illuſtrious Perſonages, who have, from time to time, thought it a kind of Duty to become Obſervers, and conſequently Admirers of this ſtupendious Sphere of primary Bodies, and diligent Enquirers into the general Laws and Principles of Nature, who can avoid being filled with a kind of enthuſiaſtic Ambition, to be acknowledged one of the Number, who, as it were, by thus adding his Atom to the Whole, humbly endeavours to contribute towards the due Adoration of its great and divine Author.

I judge it will be quite unneceſſary to ſay any thing about the Order of the Work, ſince that would be only a Repetition of the Table of Contents, to which the Reader is referred, as to the propereſt Account that can here be given.

† Seven Emperors, nine Kings, and as many ſovereign Princes. *Charlemagne* wrote *Ephemerides,* and named the Months and Winds in *High Dutch,* 770. *Rich.* II. &c.

T H E

THE
CONTENTS.

a　　　　　　　　DIREC-

DIRECTIONS for placing the PLATES.

Some of the Principal ERRATA.

Page	Line	the Words	Read.
2	ult.	to ceafe rela	ceafing to relate
4	3	Phænomen⟨	Phænomena
16	15	incomfible	incomprehenfible
21	12	comprehenc	comprehending
33	28	compared	is compared
34	37	form	from
43	20	volving	revolving
49	24	immoveabl⟨	moveable
61	19	much	much as
62	28	XXIII.	XXI.
65	4	where	any where
67	15	alfo	all fo
69	29	one	our

Plate X. read the Characters of the Planets in this Order ♃ ☿ ♄ ♂ ♀

THE

A LIST OF THE SUBSCRIBERS.

A.

LORD Anſon.
Hon. Mr. Archer.
Charles Ambler, Eſq;

B.

Duke of Beaufort.
Duke of Bedford.
Dutcheſs of Beaufort.
Lord Berkely, of Straton.
Miles Barne, Eſq;
Lancelot Barton, Eſq;
Hon. Antoine Bentinck.
Hon. John Bentinck.
Norbone Berkely, Eſq;
John Brown, Eſq;
——— Blaman, Eſq;
Thomas Brand, Eſq;
J. Bevis, M. D.
Rev. T. Bonney, A. M.

C.

Counteſs of Cunengeſby.
Lord Cornwallis.
Lady Cornwallis.
Edward Cave, Eſq;
John Chamock, Eſq;
Hon. and Rev. Dr. Cowper.
Mr. Richard Chad.
Mr. Henry Chapell.
Iſ. Colepepper.
Mr. George Conyers.

D.

Rev. John Dealtary, A. M.
Mr. Samuel Dent.

F.

Charles Fitzrea Scudamore, Eſq;
Kean Fitzgerald, Eſq;
Thomas Fonnerau, Eſq;
Robert Rakes Fulthorpe, Eſq;
Mr. Samuel Farrant.
Mr. Paul Fourdrinier.

G.

Marchioneſs Grey.
Lord Glenorchy.
Francis Godolphin, Eſq;
Roger Gale, Eſq;
James Gibbon, Eſq;
Ralph Goward, Eſq;
Ralph Gowland, Eſq;
Ralph Gowland, Junior, Eſq;
Dr. Gregory.
Dr. Griffith.
Rev. John Griffith, A. M.
Rev. Middlemore Griffith.

H.

Lord Hardwick, Lord High Chancellor of Great-Britain.
Hon. James Hamilton.
Mr. Thomas Heath.
Mr. Thomas Holt.
John Hughes, Eſq;

Earl

A LIST of the SUBSCRIBERS.

I.
Earl of Jerfey.
Richard Jackfon, *Efq;*
Rev. Mr. Jones.

K.
———— Knowles, *Efq;*
Dr. Kendrick.
Mrs. Kennon, 4.

L.
Lady Vicountefs Limerick.
Sir William Lee, *Bart.*
William Lefter, *Efq;*
Rev. Dr. Long, *Mafter of* Pembroke-hall, Cambridge.
William Lloyd, *Efq;*
Mr. Andrew Lawrence.

M.
R. J. Mead, *M. D.*
Richard Meyrick, *M. D.*
Owen Meyrick, *Efq;*
Pierce Meyrick, *Efq;*

N.
Duke of Norfolk.
Lord North.
Lord Bifhop of Norwich.
Richard Nicholls, *Efq;*
Mrs. Norfa.

P.
Duke of Portland.
Earl of Pembroke, *&c.* 2
Countefs of Pembroke, *&c.*
Lady Palmerfton.
Robert Money Penny, *Efq;*
Sir Francis Pool.
Sir John Pool.
John Probyn, *Efq;*
Rev. Mr. Pierce.

Mr. Dominick Pile.
Mr. Powel, *of* Cambridge.

R.
Dutchefs of Richmond, *&c. &c.*
James Ralph, *Efq;*
Allan Ramfey, *Efq;*
William Read, *Efq;* 2.
Henry Reveley, *Efq;*
William Reveley, *Efq;*

S.
Sir George Savile.
———— Serle, *Efq;*
Rev. Dr. Smith, *Mafter of* Trinity College, Cambridge,
Mifs Stonehoufe.
William Symonds, *Efq;*
Mr. James Scot.
Mr. James Stephens.

T.
Lord Vifcount Townfhend.
John Temple, *Efq;*
James Theobald, *Efq;*
Charles Townfhend, *Efq;*
Mrs. Mary Trevor.
Mr. James Thornton.

V.
Lord Vifcount Villiers.

W.
Lady Frances Williams.
Mifs Williams.
Mifs Charlotta Williams.
Rev. Thomas White, *A. M.*
———— White, *Efq;*
Charles Louis Wiedmarkter, *Efq;*
Mr. Ward.

Y.
Hon. Philip York.
Dr. Arthur Young, *Preb. of* Cant.

LETTER the FIRST.

Opinions of the most eminent Authors whose Sentiments on the following Subject have been published in their Works.

SIR,

REFLECTING upon the agreeable Conversation of our last Meeting, which you may remember chiefly turned upon the Stars, and the Nature of the planetary Bodies ; a Subject, which is generally allowed to give true Pleasure to all those who take Delight in mathematical Enquiries ; and having not a little Regard to the repeated Request in your late Letters, I have at length undertaken to explain to you, as far as I am able, my Theory of the *Universe*, and the Ideas I have form'd of the known Creation.

The Hypothesis upon which this new Astronomy is founded, and now reduced into a regular System, was the result of my Astronomical Studies * full fifteen years ago, hence I hope you will allow, I have more than observed *Horace*'s celebrated Aphorism,

Nonumque prematur in annum.

---

* The first Scheme of this Hypothesis was plann'd in the Year 1734, representing in a Section of the Creation, eighteen Feet long and one broad, several thousand Worlds and Systems, and a great Number of emblematical Figures, now in the Author's Possession, together with a Scheme of the entire Creation, completed since, nine Feet long and six broad, more fully illustrating upon the same Construction the Innumerability of Systems and Worlds.

B

The

The Subject, I have often obferved, you have liftened to with a pleafed Attention, and I am the more incouraged to explain it at large to you, as I am perfwaded you don't want to be convinced of its valuable Ufes and Importance.

I remember you have often told me, that to apply ourfelves to the Study of Nature, was the fureft and readieft Way to come at any tolerable Know-ledge of ourfelves, however difficult the Tafk might prove either in the Attempt, or the attaining it, and the lefs to be neglected, as it never fails to introduce a proper Knowledge of the DIVINE BEING, as a certain Con-fequence along with it, and fuch a Knowledge, as will naturally make every Man, who has but a tolerable Share of common Senfe, and is not a Slave to another's Reafon, without any other Evidence or Motive, in all Sta-tions, and under all Circumftances, ACT JUSTLY, LIVE CHEARFUL-LY, and DIE full of Hope in the Expectation of a happy Sequel, in Fu-turity.

> *Eternity* is written in the Skies:
> Mankind's Eternity, nor *Faith* alone;
> *Virtue* grows there ———
>
> *Dr.* YOUNG.

A learned Author on the Attributes, recommending thefe Studies as a reafonable and moral Service, fays, " Sure, it is moft becoming fuch im-" perfect Creatures as we are, to contemplate the Works of God with this " Defign, that we may difcern the Manifeftations of Wifdom in them; " and thereby excite in ourfelves thofe devout Affections, and that fu-" perlative Refpect, which is the very Effence of Praife."

> Who turns his Eye, *on Nature's Midnight Face,*
> *But muft enquire* ——— what Hand behind the Scene,
> What ARM ALMIGHTY, put thefe wheeling Globes
> In Motion, and wound up the vaft Machine?

The enchanting Idea *Milton* had of the Subjects of Aftronomy (whofe truly fublime Way of thinking and writing perhaps was never fo nearly equalled, or attempted before this Reverend Author's *Night-Thoughts,* ap-pear'd is finely fhewn in the Eighth Book of his *Paradife Loft,* where he makes his *Adam,* fo earneftly attentive to the Angel *Gabriel,* as to ceafe relating the Myfteries of Creation.

The

The Angel ended, and his *Adam*'s Ear
So charming left his Voice, that he awhile
Thought him ftill fpeaking; ftill ftood fix'd to hear.

Milton's own Ideas of the Univerfe too, which no doubt he had ga-
thered from aftronomical Authors, and had reconciled himfelf to, we are
fully made acquainted with in the fame Book, where the Arch-angel fays,
in anfwer to *Adam*'s Enquiries.

—— Other Suns perhaps
With their attendant Moons thou wilt defcry
Communicating Male and Female Light,
Which two great Sexes animate the World,
Stor'd in each Orb, perhaps with fome that live:
For fuch vaft Room in Nature, unpoffeft
By living Soul, defert and defolate,
Only to fhine, yet fcarce to contribute
Each Orb a Glimpfe of Light, convey'd fo far
Down to this habitable, which returns
Light back to them, is obvious to Difpute.

But before I prefume to plan my own Difcoveries and Conjectures into a
Theory, both in Juftice to thofe who have in fome meafure been in the
fame Way of Thinking, and alfo as a Defence of myfelf for producing fo
new an Hypothefis to the World, which otherwife (though any Apology
made to you I know will be unneceffary) may appear to too many but an
idle *Chimera* of my own. I judge it will be highly proper, by way
of ftrengthening my own Arguments, and adding more Weight to what
I fhall myfelf advance in the following Letters, to give you in this
the Opinions of the moft able Writers, whofe Works I have read
upon the Subject. I mean fo far as relates to the now general received
Notion, that the Stars are all Suns, and furrounded with planetary Bodies,
with which I fhall fet out; and fhew you, it is not a Thing merely taken
for granted, but has ever been the concurrent Notion of the Learned of all
Nations, as fhall be further fhewn, in its proper Place, and as nearly as
Poffibility will admit of, demonftrated to be Truth.

The following is an Extract from Mr. *Toland*, in his Account of the
Works of

JORDANUS BRUNO.

" The Divine Efficacy (fays this Author in his infinite Creation) cannot
" ftand idle, without the Want of Will or Power; but any Imbecillity in

" fuch

" fuch a Being argues Imperfection, and fince any finite Produce com-
" pared with Infinity is as nothing, or rather as the Beginning of Good,
" it muftbe no lefs idle, and invidious in producing a finite Effect, than in
" producing none at all.

" Hence, as all Finites, fingly confidered, are but as Commencements
" of fomething more to be expected.

" Omnipotence, in making the Creation finite, will appear to be no
" lefs blameable for not being willing, than for not being able, to make it
" otherwife; *i. e.* infinite, as being an infinite Agent upon a finite Subject,
" which is repugnant to Reafon."

It follows then that, Creation muft be not only extenfively, but inten-
fively indefinite, and beyond the Reach of the human Underftanding to
comprehend; and that the one is as neceffary as the other, *i. e.* an infinite Expanfe is as reconcileable to our Reafon, as infinite Parts are to our
Senfes.

All the Attributes of the Divine Being are, as any one of them, incom-
fible to his Creatures; why fhould our Imagination then be fuppofed to
extend beyond the divine Activity?

" Thus, adds the above Author, the Excellency of God is adequately
" magnified, and the Grandeur of his Empire made manifeft; he is not
" glorified in one, but in numberlefs Suns; not in one Earth, or in one
" World, but in ten thoufand thoufand of infinite Globes."

An infinite Reprefentation of an infinite Original, and a Spectacle befit-
ting the Excellency and Eminence of him, that can neither be fully conceived, imagined, or comprehended.

> What read we here? th'Exiftence of a GOD?
> Yes, and of other Beings, Man above,
> Natives of Æther! Sons of higher Climes!

Dr. YOUNG.

" If the Exiftence of this one World be good or convenient, it is not
" lefs good or convenient that there be infinite others like it.

" The infinite efficient Caufe would be abfolutely defective, without an
" infinite Effect; and befides, by conceiving the Infinity of the Univerfe
" and innumerable Beings, the Underftanding refts fatisfied, and is recon-
" ciled with the Idea of an Eternity; whereas, by afferting the contrary,
" it is unavoidably plunged into innumerable Difficulties, and unfolvable
" Inconveniencies, Paradoxes, and Abfurdities.

Again, fays the fame Writer, " Did we but confider and comprehend
" all this, oh! to what much further Confiderations and Comprehenfions
" fhoud

" fhould we be carried! as we might be fure to obtain that Happinefs
" by virtue of this Science, which *in other Sciences is fought after in vain.*

> This Profpect vaft, what is it? weigh'd aright,
> 'Tis Nature's Syftem of Divinity,
> And every Student of the Night infpires.
>
> Dr. YOUNG.

> 'Tis elder Scripture, writ by GOD's own Hand;
> Scripture authentic! uncorrupt by Man.

" This then is that Philofophy, which opens the Senfes, which fatisfies
" the Mind, which enlarges the Underftanding, and which leads Man-
" kind to the only true Beatitude, whereof they are capable according to
" their natural State and Conftitution ; for it frees us from the follicitous
" Purfuit of Pleafure, and from the anxious Apprehenfions of Pain, mak-
" ing us to enjoy the good Things of the prefent Hour, and not to fear
" more, than we hope from the future ; fince that fame Providence, or
" Fate, or Fortune, which caufes the Viciffitudes of our particular Being,
" will not let us know more of the one, than we are ignorant of the
" other."

And farther, " From thefe Contemplations, if we do but rightly confider,
" it will follow, that we ought never to be difpirited by any ftrange Ac-
" cidents, through Excefs of Fear or Pain, nor ever be elated by any prof-
" perous Event, through Excefs of Hope or Pleafure ; whence we have
" the Path to true Morality, and following it, we fhall of courfe become
" the magnanimous Defpifers of what Men of weak Minds fondly
" Efteem, and be wife Judges of the Hiftory of Nature, which would be
" written in our Minds, and confequently be chearful and ftrict Execu-
" tioners of the divine Laws, which would thus be ingraved in the Cen-
" ter of our Hearts. Seeking, as it were, in ourfelves, an Approbation of
" our own Action, which alone is capable of true Content and Happi-
" nefs."

CHRISTOPHER HUYGENS,

To whom the World is much indebted for many curious Inventions, and
Difcoveries, fays in his *Planetary Worlds,* " I muft be of the fame
" Opinion with all the great Philofophers of our Age, that the
" Sun is of the fame Nature with the fix'd Stars ; and this will give us a
" greater

+ The Pendulum Clock ; the firft Difcovery of *Jupiter's* Satellites, and *Saturn's* Ring.

" greater Idea of the World than all other Opinions can. For then
" why may not every one of thefe Stars, or Suns, have as great a Retinue,
" as our Sun, of Planets, with their Moons to wait upon them? Nay,
" there is a manifeft Reafon why they fhould; for, if we imagine our-
" felves placed at an equal Diftance from the Sun and fix'd Stars, we
" fhould then perceive no Difference at all betwixt them.

" Why then may we not make ufe of the fame Judgment that we
" would in that Cafe; and conclude, that our Star has no better Atten-
" dance than the others? So that what we allowed the Planets upon the
" Account of our enjoying it, we muft likewife grant to all thofe Planets
" that furround that prodigious Number of Suns. They muft have their
" Plants and Animals, nay, their rational Creatures too, and thofe as great
" Admirers and as diligent Obfervers of the Heavens as ourfelves; and
" muft confequently enjoy whatever is fubfervient to, and requifite for
" fuch Knowledge.

" What a wonderful and amazing Scheme have we here of the mag-
" nificent Vaftnefs of the Univerfe! So many Suns, fo many Earths, and
" every one of them ftock'd with fo many Herbs, Trees, and Animals,
" and adorned with fo many Seas and Mountains! And how muft our
" Wonder and Admiration be increafed, when we confider the prodi-
" gious Diftance and Multitude of the Stars?"

The Opinion of Sir ISAAC NEWTON.

This great Author, in his grand *Scholia* to the *Principia*, fays: — " The
" moft beautiful Syftem of the Sun, Planets, and Comets, could only pro-
" ceed from the Counfel and Dominion of an intelligent and powerful
" Being: And if the fix'd Stars are the Centers of other like Syftems, thefe,
" being form'd by the like wife Counfel, muft be all fubject to the Do-
" minion of One; efpecially, fince the Light of the fix'd Stars is of the
" fame Nature with the Light of the Sun, and from every Syftem Light
" paffes into all the other Syftems. And leaft the Syftems of the fix'd
" Stars fhould by their Gravity fall mutually on each other, he (the Di-
" vine Being) hath placed thofe Syftems at immenfe Diftances from one
" another."

The

The Opinion of Dr. DERHAM, *in his* Aftro-Theology.

" The new Syftem, fays he, fuppofeth there are many other Syf-
" tems of Suns and Planets, befides that, in which we have our
" Refidence ; namely, that every fix'd Star is a Sun, and incompaffed
" with a Syftem of Planets, both primary and fecondary, as well as ours.

" Thefe feveral Syftems of the fixed Stars, as they are at a great and
" fufficient Diftance from the Sun and us ; fo they are imagined to be at
" as due, and regular Diftances from one another : By which means it is
" that thofe Multitudes of fixed Stars appear to us of different Magnitudes,
" the neareft to us large ; thofe farther and farther, lefs and lefs ; and
" that fome, if not all of thofe vaft Globes of the Univerfe, have a Mo-
" tion, is manifeft to our Sight, and may eafily be concluded of all, from
" the conftant Similitude and Confent that the Works of Nature have
" with one another."

To this we may add, that this Syftem of the Univerfe, as it is phyfi-
cally demonftrable, is far the moft rational and probable of any. *Becaufe,*

" It is far the moft magnificent of any, and worthy of an infinite
" CREATOR, whofe *Power* and *Wifdom,* as they are without Bounds and
" Meafure, fo may they, in all Probability, exert themfelves in the Creation
" of many Syftems as well as one. And as Myriads of Syftems are more
" for the *Glory* of GOD, and more demonftrate his *Attributes* than one ;
" fo it is no lefs probable than poffible, there may be many befides this
" which we have the Privilege of living in." And as the ftrongeft Confirmation of this, " we fee it is really fo, as far as it is poffible it can be
" difcerned by us, at fuch immenfe Diftances as thofe Syftems of the fixed
" Stars are from us ; and we cannot reafonably expect more."

" Since the Sun and fix'd Stars, fays Dr. *Gregory,* are the only great
" Bodies of the Univerfe that have any native Light, they are juftly
" efteemed by Philofophers to be of the fame Kind, and defigned for the
" fame Ufes ; and it is the Effect of a Man's Temper that fets a greater
" Value upon his own Things than he ought, that makes him judge
" the Sun to be the biggeft of them all."

That, as an elegant * Writer obferves, which we call the Morning, or
the Evening Star, is, in reality, a *Planetary World* ; which, with the four
others, that fo wonderfully, as *Milton* expreffes it, " vary their myftick
" Dance, are in themfelves dark Bodies, and fhine only by Reflection ;
" have Fields and Seas, and Skies of their own ; are furnifhed with all
" Accommodations for animal Subfiftence, and are fuppofed to be the

Abodes

* Contemplations on the ftarry Heavens.

" Abodes of intellectual Life. Again, The Sun, with all its attendent Planets
" is but a very little Part of the grand Machine of the Universe. Every
" Star— is really a vast Globe, like the Sun, in Size and in Glory, no less
" spacious, no less luminous, than the radiant Source of our Day ; so that
" every Star is the Center of a magnificent System, has a Retinue of
" Worlds irradiated by its Beams, and revolves round its active Influence;
" all which are lost to our Sight in immeasurable Tracts of Æther.

" Could we, says the same Author, wing our Way to the highest ap-
" parent Star — we should there see other Skies expanded, other Suns,
" that distribute their inexhaustible Beams of Day ; other Stars, that gild
" the alternate Night ; and other perhaps nobler Systems established ;
" established in unknown Profusion, through the boundless Dimensions
" of Space. Nor does the Dominion of the great Sovereign end *there*,
" even at the End of this vast Tour, we should find ourselves advanced
" no farther than the Frontiers of Creation ; arrived only at the Suburbs
" of the great *Jehovah*'s Kingdom."

> O for a Telescope his Throne to reach !
> Tell me ye Learn'd on Earth ! or Blest above !
> Ye searching, ye *Newtonian* Angels ! tell,
> Where your great Masters Orb ? His Planets where ?
> Those conscious Satellites, those Morning Stars,
> First-born of *Deity* from central Love.
>
>   Dr. YOUNG.

Many other Authorities might be produced from Writers of great Repute, were it necessary to trouble you with them † ; but I believe those
above will be abundantly sufficient for the present Purpose, if even an
Apology were wanting for my own Conjectures. I shall therefore conclude this Letter with the following Passage out of *Pope's universal Prayer*,
and in my next shall proceed in the Work I have undertaken.

> Yet not to Earth's contracted Span,
>   Thy Goodness let me bound ;
> Or think thee Lord alone of Man,
>   When thousand Worlds are round.
>
>   *I am*, &c.

                                    LETTER

† Particularly from *Fontenelle*, &c.

LETTER the SECOND.

Concerning the Nature of Mathematical Certainty, and the various Degrees of Moral Probability proper for Conjecture.

S I R,

YOU know how much I am an Enemy to the taking of any thing for granted, merely because a Person of reputed Judgment, has been heard to say, *it absolutely is so*; an *Ipse dixit*, and implicit Faith in some Cases, may be both necessary and useful; but here, in Astronomy, I mean, every Man's Reason, by the Help of a very little Mathematicks, is able to bring wonderful Truths to Light without them; and Truths not only of the highest Importance to every Individual, but of a great and common Consequence to all Mankind: And as such, in all Ages of the World, have been judged worthy to be enquired into, by the best and wisest of Philosophers.

You are likewise very sensible how far the human Understanding is even at the best, from being infallible, and don't want to be told, how difficult it is in a Subject of this Nature to arrive at any tolerable Degree of Certainty, which before the Days of the sagacious *Euclid*, and the penetrating *Archimedes*, was a Thing not to be expected. And many things which were then but barely Objects of Conjecture and Probability, have since been demonstrated to be infallibly true. Time and Observation will undoubtedly, at last, discover every thing to us necessary to our Natures, and proper for us to know. As a Proof of which, we see human Wisdom daily increases; and while a Capacity continues to make ourselves still more acquainted with the manifest Wisdom and Power of GOD in the Works of his Creation, who is to tell us where to stop our Enquiries? Or who is so impious to set Bounds to a Science, which so evidently spreads through all Infinity, the Attributes of God, and an eternal Basis for future Hope?

This Branch, or rather Body of Astronomy, I believe you will find to be quite new; and though evident Truths, are the principal Thing to be regarded in it, yet as being in its infant State, where lineal Demon-

C   stration

ſtration fails, as in ſome Caſes it cannot be otherwiſe, I hope you will give me Leave to make uſe of a weaker Way of Reaſoning, to convince you of the Point in Diſpute, I mean of that by the Analogy of known and natural Things.

I ſhall be extremely unwilling to affirm any thing for a *Fact*, or Truth, without hearing, if not the real Evidence, at leaſt a plauſible Reaſon, next to a Conviction, or moral Certainty, along with it; and therefore I will here endeavour to explain to you what I mean by moral Certainty and alſo by mathematical Proof.

Mathematical Proof, or Certainty, proper for Conjectures, may, to almoſt every Capacity, be illuſtrated as follows

Suppoſe you had accidentally found a very ſmall Part of a viſibly broken Medallion, with nothing more expreſs upon it, than what is repreſented at *Fig.* 1. *Plate* I. a Perſon totally unacquainted with the mathematical Sciences, we may naturally conclude, would not be able to make any thing of it, or in the leaſt comprehend what it originally was, or meant; but if an Aſtronomer ſhould chance to ſee it, who of courſe we are to ſuppoſe knew the Order and Proportion of the planetary Orbits, he would immediately conclude, and with great Probability, on the Side of his Conjectures, that it might be Part of a Medal repreſenting the Solar Syſtem. In ſuch a Caſe may we not very naturally ſuppoſe he would reaſon thus?

The Arches A and B ſeem to be Portions of the reſpective Orbits of *Saturn* and *Jupiter*, and what may lead us to believe, that they are really ſo, and Part of the Solar Syſtem, is the oblique Curve C, which looks not unlike the Trajectory of a Comet.

This ſurely would be far from an irrational Conjecture, and conſequently in ſome Degree probable: But this is not ſufficient you'll ſay; To prove it we muſt have farther recourſe to the Mathematicks, and a Mathematician would immediately thus demonſtrate it to be true.

Firſt, by compleating the Circles geometrically from the fourth Book of *Euclid*, by the Aſſiſtance of any three Points E. F. G. the original Figure will be reſtored, as at *Fig.* 2. And ſecondly, by aſſuming any two Points, as F, E in the Curve C, if admitted a Parabola, by a well-known Problem in Conic Sections the Heliocentric Portion X. Y. Z. will eaſily be projected and ſhewn, as in *Fig.* 3. Laſtly, join this in Poſition to the former, and it will juſtly ſupply the Orbit, or Path of ſome one of the Comets; and if required, even what Comet may be diſcovered by comparing the Perihelion Diſtance Y. S. with their general Elements or Theories, in Dr. *Hally*'s *Synopſis* of the Motion of theſe Bodies. And if a farther Confirmation of the Truth of theſe Conjectures were wanting,

the

PLATE.I.

Fig. 1.

Fig. 2

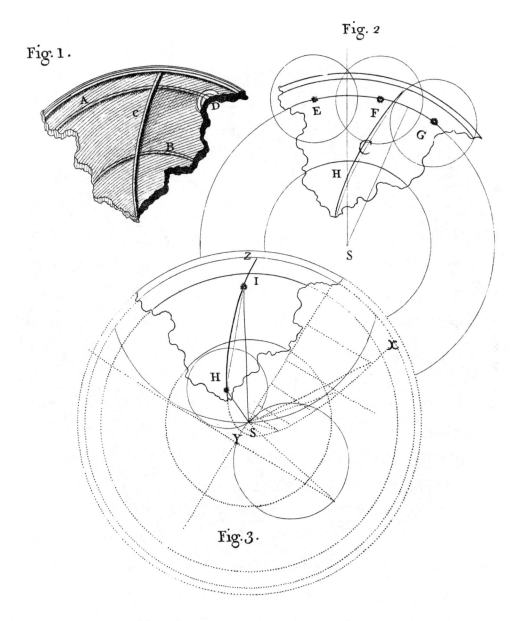

Fig. 3.

PLATE II.

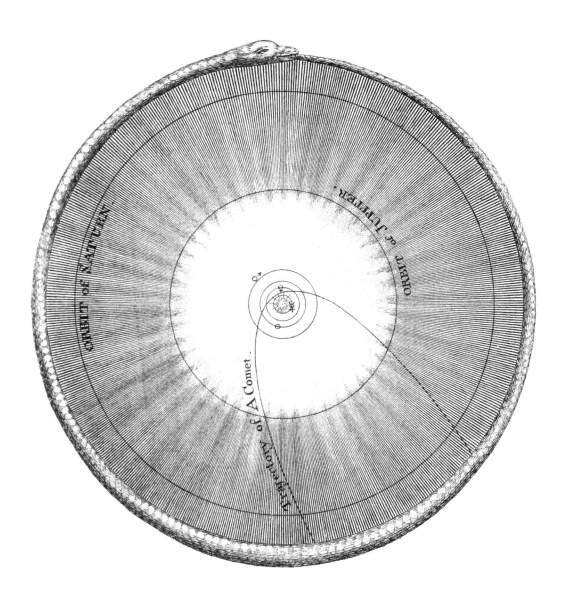

the fmall concentric Circles at D would now be allowed beyond a Contradiction, to reprefent the fecondary Orbits of *Saturn*; and thus the
firft Prefumption being carried thro' feveral corroborating Degrees of Probability, almoft paft a Difpute, would become a mathematical Certainty;
and the above imperfect Piece of Medallion, would evidently appear beyond a Contradiction to be Part of a Reprefentation of the faid folar Syftem,
and fuch as is fhewn in *Plate* II. Q. E. D. Thus in many Cafes, it often hap-
pens, that from a very fmall Part of *orbicular Things*, we are able to determine the Form and Direction of the Whole: And hence you may
conceive it no very difficult Tafk to a Mathematician, to defcribe the Orbits of all the Planets in the folar Syftem, though he had never obferved
them but in one and the fame Sign of the *Zodiack*; thus far I have thought
it would not be amifs to explain to you the Nature of thofe Steps, by
which we arrive at moral Certainty, and where the Subject will admit of
it, Mathematical Conviction, which will not a little contribute to ftreng-
then many of the Arguments hereafter made ufe of, and in fome Degree
ferve to fupply the Place of Proof, where infallible Demonftration cannot from the Nature of the Thing be difcovered.

But befides the indifputable Principles of Geometry, the univerfal
Law of Analogy and Similitude of things, have a Privilege to affift us,
in Conjectures relating to the heavenly Bodies, and though not of equal
Force with the former, is often as conclufive as the Subject requires. This
fort of probable Evidence (as Dr. *Butler* obferves,) is effentially diftinguifhed from " Demonftrative by this, that it admits of Degrees; and
" of all Variety of them, from the higheft moral Certainty to the very
" loweft Prefumption; and that which chiefly conftitutes Probability, is
" expreffed in the Word *Likely*, or Natural Likenefs, as to State or Being."
This general Way of arguing, I think, is allowed to be evidently natural,
juft and conclufive, and unqueftionably to have its Weight in various Degrees, towards determining our Judgment: For Inftance, fhould any ignorant Perfon, endowed with rational Principles, cut open a *Pomegranate*
of the natural Growth of *England*, and finding it full of fmall Globules, or
Kernels, upon being prefented with an every way fimilar Fruit, faid to be
the Produce of *Italy*, doubt of its being of the fame Nature, and com-
pofed of like globular Seeds within; here indeed would be no mathematical Evidence to affift the Judgment, the Object of Proof being invifible,
but fure from the external Similitude, the ftrongeft Probability of their
being alfo internally the fame. Again,

Is it natural to fuppofe, that the firft Perfon who found a *Lark*'s Neft,
and in it feveral of the Female's Eggs, fhould have any Apprehenfions of
finding none in the *Nightingale*'s, only becaufe he had never feen one be-

fore,

fore, I believe the most illiterate Person of the earliest Ages, who had Curiosity enough for such a Search, would be greatly disappointed in such a Case, and far from concluding that the *Nightingale* had none. Farther, should any one who had seen several Sorts of Fish taken out of the River *Thames*, or out of the *Nyle*, have any sort of Suspicion that he should find no such Creatures in the *Seine* or the *Ganges*, though it should be allowed that he had never seen any such Creatures that were known to come from thence. Ocular Demonstration, in such a Case, would sure be unnecessary, and an Evidence of the first, I believe would be abundantly sufficient to convince us of what we ought to look for at least in the last: But then the Fishes of different Seas, and of Rivers are not of the same Species you'll say; but as it were infinitely diversified through all the aqueous World, this is, and must be granted, and alike Variety of *Species* must also be granted, in the former Case of the Birds: But no Objection can possibly arise from any such Diversity, since we don't pretend to say, nor is it at all necessary, that the Beings in the sidereal Planets should be every where the same with these of our solar System, a Variety must every where be admitted, and will always be admired, where the Work is Nature's, and the Design GOD's.

All then that I here pretend to argue for, is a Universality of rational Creatures to people Infinity, or rather such Parts of the Creation, as from the Analogy and Nature of Things, we judge to be habitable Seats for Beings, not unlike the mortal human.

Every Animal, and every Vegetable, that, as it were, naturally exists by the Virtues, Properties, or Laws of the mineral Kingdom, has something of a secondary Nature, depending upon it as a Principle; and to say that the Stars, which are a certain visible sort of Cotemporaries in Space with the Sun, have no like planetary Bodies with ours moving round them, because we cannot possibly see them, is no less absurd and ridiculous, than to argue, that we can have no Reason to expect to find, in the proper Season, Grapes upon every Vine — Figs upon every Tree — Roses upon every Bush — only because some of them are at such a Distance, that neither Rose, Fig or Grape, can be discovered by the Eye.

This sort of Reasoning, though some perhaps may neglect it, I am perswaded you will look upon as abundantly sufficient for Things out of the Reach of Science to determine; and that the collective Body of Stars have not been discovered, to be together a proper Subject for such Conjectures before, can surely only proceed from the Want of Time, necessary to compleat the Observations proper for a Foundation to build such an Hypothesis, or Theory upon. This is the great Article in which the Moderns have so much, and ever will have, an Advantage over the Antients. And hence it will appear, That

The

The Improvements and Difcoveries of latter Ages are not at all owing to the greater Capacity of the Moderns, but from the Advantages receiv-ed, or arifing from the Inventions and Progrefs made by the Ancients. We at firft in a manner walked by their Leading ftrings, and though many of them now are broke, or ufelefs, none can deny, but that formerly they were of great Advantage in promoting and directing philofophical En-quiries.

In an Affembly of the moft eminent Men of all Ages, if we may fup-pofe fuch a Conference amongft the illuftrious Dead, on Purpofe to deliver their feveral Sentiments familiarly together, on the moft interefting Sub-jects of natural Knowledge, who would not lament the Difadvantages, poor old *Thales*, an *Hipparchus*, or a *Ptolomy*, would lie under, who had no-thing but the Eye of Reafon to direct them, in Oppofition to the Judg-ment of a *Brahe*, or a *Galilæus*, who reaped fo much Benefit from their compound Opticks? But on the other hand, perhaps if the folar Syftem, was the Topic of Difcourfe, a * *Pythagorean* might very pertinently fay to a *Newtonian*, " You have not gone much farther in the Light with our " Direction, than we did in the Dark alone ; for you are ftill roving " round the fame Circles." Much might be faid upon this Head ; but I believe it would be a difficult Matter to do Juftice to all Parties: So here I intend to leave them, only muft obferve, that Pofterity will always have the Advantage over their Predeceffors ; and that After-ages, in all Pro-bability, will reap fo great a Benefit from the Invention and Improve-ment of Fluxions, that fcarce any thing, which is the immediate Ob-ject of fuch Enquiry, will long lie concealed from a true mathematical Genius.

For this, in which he has furpaffed all the Antients, and greatly advanced the philofophical Sciences, the World is indebted to Sir *Ifaac Newton*.

But as many of his Difcoveries, fuch as relate particularly to the Laws of the planetary Syftem, are but as fo many Confirmations of the Con-jectures and Imaginations of Aftronomers and Philofophers before him, it perhaps will not be amifs to acquaint you a little with the Aftronomy of the Antients concerning the Univerfe. And before I proceed to thofe of my own, fhew you in the firft Place how far their Speculations in the vi-fible Creation have been carried ; and with thefe I fhall conclude this pre-paratory Epiftle.

The Univerfe, or mundane Space, by which the Antients comprehend all Creation, has, from time to time, according to the Progrefs of Science, come under a fort of Neceffity of being varioufly modell'd agreeable to the

Opinion

---

* The true Syftem of the Planets have been difcovered above two thoufand Years.

Opinion of the feveral Authors, who have judged themfelves wife enough to write upon it with a mathematical Foundation: And the cofmical Syftem, by which is meant the Co-ordination of its conftituent Parts has undergone almoft as many Changes as its Elements are even capable of ; every Age of the World, as Knowledge has increafed, either from improved Imagination, or repeated Obfervations, producing fomething new concerning it.

MILTON, no doubt, had all this Diverfity of Opinions in View, as appears from his fuppofed Pre-knowledge of *Raphael*, in the following Paffage, *Book.* VIII.

> Hereafter, when they come to model Heaven,
> And calculate the Stars, how they will weild
> The mighty Frame ! how build, unbuild, contrive
> To fave Appearances, how gird the Sphere
> With centric and eccentric fcribbl'd o'er ;
> Cycle, and Epicycle, Orb in Orb.

But the following Synopfis, I believe, will abundantly convince you that from certain Obfervations only, we ought to form all our Notions of it, if we either hope to arrive at Truth, or expect our Ideas fhould be fupported by Reafon.

ARISTOTLE was of Opinion, that the Univerfe, or Heaven, was all one World, and St. CHRYSOSTOM, TERTULLIAN, St. BONAVENTURE, TYCHO BRAHE, LONGOMONTANUS, KEPLER, BULIALDUS and TELLEZ, were of an united Opinion, that this one Heaven, or Univerfe, was all fidereal and fluid. But AEGIDIUS, HURTADUS, CISALPINUS, and AVERSA, believing the fame Heaven with them to be all one World, and that fidereal, yet on the contrary held it to be folid.

CLEMENS, ACACIUS, THEODORET, ANASTASIUS, SYNAITA, PROCOPIUS, SUIDUS, S. BRUNO, and CLAUDIANUS MAMERTUS, fuppofed the univerfal mundane Space as divided into two Heavens, namely,

The Empyræum created the firft Day,

And the Firmament created the fecond Day.

Two Heavens were alfo held by JUSTIN MARTYR, the one fidereal, and the other aerial. The firft fuppofed by St. GREGORY NYSSENE, to be that of the fixed Stars, and the laft, that of the Planets. But *Maftrius* and *Bellutus*, though agreeing in the Number of Heavens, call one the *Primum Mobile*, and the other, the Starry Heaven.

Farther,

Farther, St. BASIL, St. AMBROSE, DAMASCENE, CASSIODORUS, GE-NEBRARDUS, SUAREZ, TANNERUS, HURTADUS, OVIEDUS, TELLEZ, and BORRUS, diftinguifhed the Univerfe as divided into three Portions, or Heavens.

		Or, as Cajetan.		*Tho. Aquinas.*	
The firft called the Empyræum,			Watery,		
The fecond fuppofed Sidereal,			Sidereal,		Watery,
And the laft of all, Aerial.			Aerial,		Sidereal.

Again, St. *Athanafius* adds to thofe of the fix'd Stars, the Planets, and the Air, that of the *Empyræum*, and makes in all four Heavens.

But as the Number of the Heavens thus increafes, and will become fubdivided in the fubfequent Account of them, to give you a better Idea of the Order of thefe celeftial Portions of the mundane Space, it will not be amifs to form what remains of them into regular Sections of their proper Spheres and Syftems.

See *Plate* III. in which Figure, the firft reprefents a Section of the cofmical Theory of *Oviedus* and *Ricciolus:* Both confifting of five Heavens, *viz.*

By Oviedus, fidereal and folid.			*By Ricciolus, fider. and fluid.*	
The fixed Stars,	A		*Empyræum*, - - G	
Saturn, - - - - - - B			The Water, - - F	
Jupiter, - - - - - C			The fixed Stars, A	
Sol, with ♂, ☿ and ♀ included D			The Planets, - H	
The Moon. - - - - E			The Air. - - I	

Fig. II. reprefents that of venerable *Bede* and *Rabanus*, *viz.* of Seven Heavens.

And according to *Bede* compofed of		But by *Rabanus*,
The Air, - - — - - P		The Atmofphere,
The Æther - - - - O		The upper Air,
Olympus, - - - - - N		The inferior Fire,
The Element of Fire, - - - M		The fuperior Fire,
The Firmament, - - - - A		Sphere of the fixed Stars,
The Angelical Region, - - L		The Chryftalline Heaven,
Realm of the Trinity. - - K		The *Empyræum*.

Fig.

Fig. III. Reprefents the Hypothefes of *Eudoxus, Plato, Calippus, Cicero, Riccius, Philo, Remigius, Aben-Ezra, Carthufianus, Lyranus, Tojlatus, Brugenfis, Orontius, Cremoninus, Philalethæus, Amicus,* and *Ruvius*; alfo the *Babylonians* and *Egyptians.*

Confifting of Eight Heavens,

All Sidereal, *viz.* The Sphere of the fix'd Stars, and thofe of the Seven Planets.

Fig. IV. is that of *Macrobius, Haly Alpetragius, Rabbi-Jofue, Rabbi Moyfes, Scotus, Abraham Zagutus, Sacrobofcus, Claromontius, Avigra,* and *Arraiga.*

All of Nine Heavens,

Comprehend a *Primum Mobile* Q, or, according to *Arriaga,* a folid *Empyræum.* The Sphere, of fixed Stars A, and the feven Regions of the folar Planets.

Fig. V. is that of the great *Alphonfus, Fernelius, Regiomontanus, Amicus, Maurolycus* and *Langius*; alfo of *Azabel, Thebit,* and *Ifaac Ifraelita*; and likewife of *Gulielmus Parifienfis,* and *Johannes Antonius Delphinus.*

Confifting of Ten Heavens, made up of

A *Primum Mobile* — — —   S *Empyræum.*
A Sphere of *Tripidation* in Longitude —   R *Primum Mobile.*
The Sphere of the fixed Stars — —   A
And thofe of the feven folar Planets within.

Note, Some Authors place the Sphere of *Tripidation* in Longitude below that of the *Aplain,* or Eighth Sphere.

Laftly, Fig. VI. is the Heaven of *Petrus Alliacenfis,* the College of *Conimbra, Martinenfis,* (and fometime) of *Clavius*; and alfo *Johannes Warnerus, Leopoldus de Aufriá, Johannes Antonius Maginus*; and laftly, of *Clavius.*

In all Eleven Heavens containing,

T   A *Primum Mobile,* or, as others fay, an *Empyræum.*
V   A Sphere of Libration in Latitude.
W   A Sphere of Libration in Longitude.
A   The Sphere of the fixed Stars, and thofe of the Planets.

Thus you fee how many various Opinions have from time to time been imbraced concerning the Fabric and Formation of the vifible Univerfe; all of which are now and have long been exploded; and although at firft advanced by Men of the greateft Learning, and of the deepeft Penetration in natural Knowledge, it does not appear from any one of their Opinions, that they had any the leaft Notion of infinite Space, but as it

were

PLATE III.

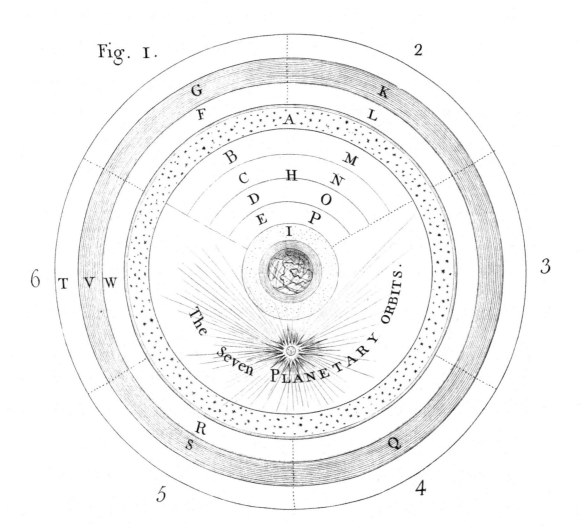

Fig. I.

The Seven PLANETARY ORBITS.

were confined the Divine BEING to their limited Notions, as one may say in an Egg-ſhell. If therefore what I ſhall hereafter advance, extend ſo far without the known Creation, that you can poſſibly conceive no Bounds to the Works of infinite Wiſdom and Power, I hope you will be in no Danger of looking upon it as more ridiculous, or abſurd, than what ſo many of the wiſeſt Men of every Age have thought proper to attempt, and have judged worthy of their Attention ſo long before me. If any thing leſs ſo, I ſhall think myſelf happy enough in having broke, or rather paſſed the narrow Limits to which the Creation has for ſo many Years been confined, in hopes of tempting Men of greater Talents to look up wards, and purſue ſo noble a Subject as far as the human Underſtanding is capable of comprehending it.

To the Opinions above might be added many more, particularly that of *Johannes Baptiſta Turrianus*, and *Fracaſtorius*, who increaſed the Number of Heavens to fourteen, *viz.* ſeven on each Side the *Aplané*.

But of this I have ſaid enough; in my next I ſhall proceed to Matter better grounded,

And am, &c.

LETTER the THIRD.

Concerning the Nature, Magnitude, and Motion of the Planetary Bodies round the Sun, &c.

SIR,

THE younger *Pliny*, if I remember right, somewhere says, that there is, or ought to be, a wide Difference betwixt writing to a Friend, and writing to the Publick: I have indeed pleased myself with the one, but am far from thinking myself qualified for the other; I must therefore rather intreat you, though perhaps you cannot possibly overlook all my Faults as an Author, to excuse them at least in the Friend, and by such kind of unlimited Indulgence, you will give me a much greater Chance to do the Subject some Justice, though I own I despair in this first Attempt, to reconcile every thing I advance to your more cool and impartial Reasoning. But to the Business:

As I have no Ambition to have the Substance of my Theory more admired by you than understood, which is too often the Case in Works of this Nature, I must beg leave to repeat to you Part of a former Discourse, which will refresh in your Ideas the principal Laws of the System of our Sun, and make you properly acquainted with such Things as are necessary to be known in the now-established Astronomy of * *Copernicus*, &c. before I proceed to any new Matter.

The

---

* NICOLAUS COPERNICUS, stiled by *Bulialdus*, *Vir absolutæ subtilitatis*, was a Native of *Thorn* in *Polish Prussia*, and Canon of the Church of *Frawenburgh*; he was Scholar to *Dominicus Maria* of *Ferrara*, to whom he was Assistant in his astronomical Observations at *Bologne*, and Professor of the Mathematicks at *Rome*, in his noble Work, *De Revolutionibus Orbium Cælestium*; he fortunately revived, happily united, and formed into an Hypothesis of his own, the several Opinions of *Philolaus*, *Heraclides Ponticus*, and *Ecphantus Pythagoreus*, *viz.* after the Opinion of *Philolaus* he made the Earth to move about the Sun, as the Center of its annual Motion; and according to *Heraclides* and *Ecphantus*, he likewise gave it a diurnal Rotation round its own Axis: Which System has withstood all Opposition; and as *Ricciolus*, (though a Dissenter from it) observes, *Per damna, per cædes, ab ipso sumit opes, animumque ferro.*

The Sun, you are not to learn, is the reputed Center of our *Planetary System*, and may remember, that the Earth on which we live, and thefe five following *Erratic Stars*, viz. SATURN, JUPITER, MARS, VENUS and MERCURY, have been demonftrated to move round him in the Order and Manner following.

Saturn is found to complete one Revolution round the Sun in twenty-nine Years, one hundred and feventy-four Days, fix Hours, and thirty-fix Minutes; at the Diftance of about feven hundred and feventy-feven Million of Miles. *Jupiter* performs a like Revolution in about eleven Years, three hundred and feventeen Days, twelve Hours, and twenty Minutes; diftant from the Sun about four hundred and twenty-four Millions of Miles. *Mars* compleats his Circuit in one Year, three hundred and twenty-one Days, twenty-three Hours, and twenty-feven Minutes; and his mean Diftance is about one hundred and twenty-three Millions of Miles.

Thefe three are called fuperior Planets, as being farther from the Sun than the Earth, and circumfcribing its Orbit.

The Earth circumambiates her Orbit in one folar Year, *viz.* in three hundred and fixty-five Days, five Hours, forty-eight Minutes, and fifty-feven Seconds; at the mean Diftance of eighty-one Million of Miles.

The Radius of *Venus*'s Orbit is about fifty-nine Millions of Miles; and that of *Mercury* nearly thirty-two Millions, *ditto*.

The Heliocentric Revolution of *Venus*, is made in two hundred and twenty-four Days, fixteen Hours, forty-nine Minutes, and twenty-feven Seconds; and that of *Mercury*, in eighty-feven Days, twenty-three Hours, fifteen Minutes, and fifty-four Seconds. Thefe two laft Planets are called inferior Ones, as being circumfcribed by the Earth.

The Diameter of the Sun being demonftrated to be nearly feven hundred and fixty-three thoufand Miles:

The proportional Magnitudes of all the above Planets will be found nearly as follows, *viz.*

The Diameter of the Globe,

Of *Mercury*	- -	4,240
Venus	- -	7,900
the Earth	- -	7,970
Mars	- -	4,440
Jupiter	- -	81,000
and *Saturn*	- -	61,000

⎫ Miles.

Thus

Thus much I have thought proper to premise, and for your immediate Inspection, have added the following Schemes, that nothing may be wanting to give a general Idea of the Order of the celestial Bodies in our own System, before I attempt to lead you through the neighbouring Regions of the Stars to the more remote Tracts of Infinity.

PLATE IV.

Is a true Delineation of the solar System, with the Trajectories of three of the principal Comets, whose Periods and Orbits have been accurately determined, and are represented in their true Proportion and Position to one another, and the Order of the Planets round the Sun, marked with their respective Characters, *viz.* ♄, for *Saturn*, ♃, *Jupiter*, ♂, *Mars*, ⊕, the Earth, ♀, *Venus*, and ☿, *Mercury*. The Scale being nearly five hundred and eighteen Millions of Miles to an Inch.

PLATE V.

Is a true Projection of the System of the known Comets; in which are represented nine of the chief Trajectories, from their *Aphelii* to their *Perihelii*, all in just Proportion and Position to the Orbits of *Saturn* and *Jupiter*, which are also represented by the two concentric Circles, supposed to be drawn round the Sun as their Center.

The Ellipsis, or Trajectory, marked A, shews the Position and Path of the Comet which appeared in the Year 1684, whose Period is supposed to be about fifty Years, and has been observed within the Region of the Planets once.

That mark'd B, is the Way of the Comet of 1682 ;
The Period conjectured to be about seventy-five Years and a half, and has been observed thrice.

C, Way of the Comet of 1337 ;
The Period about 100 Years, observed once.

D, That of the Comet of 1661 ;
The Period about 129 Years, observed twice.

E, Tract of the Comet of 1618 ;
The Period about 160 Years, observed once.

F, Way of the Comet of 1677 ;
The Period about 200 Years, observed once.

G, Way of the Comet of 1744 ;
The Period about 300 Years, observed once.

H, Way of the Comet of 1665 ;
The Period about 400 Years, observed once.

I, Way of the Comet of 1680 ;
The Period about 575 Years, observed thrice.

The

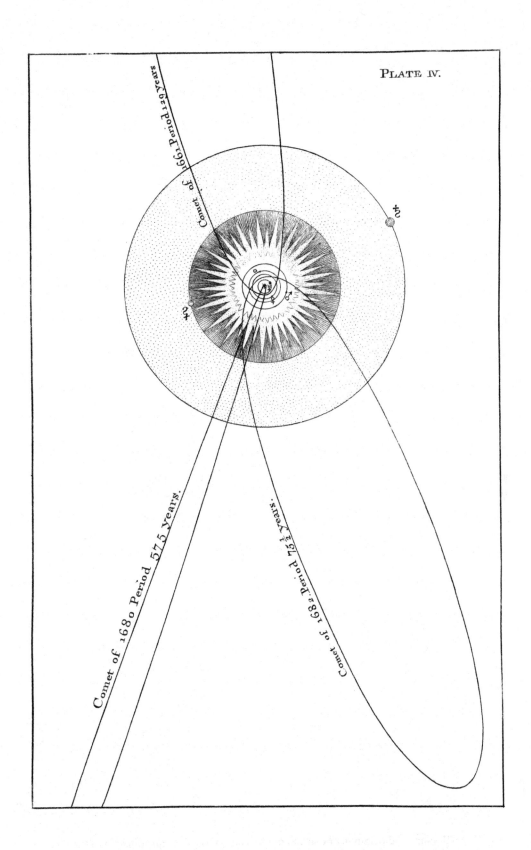

PLATE IV.

Comet of 1661. Period 129 years

Comet of 1680 Period 575 years.

Comet of 1682. Period 75½ Years.

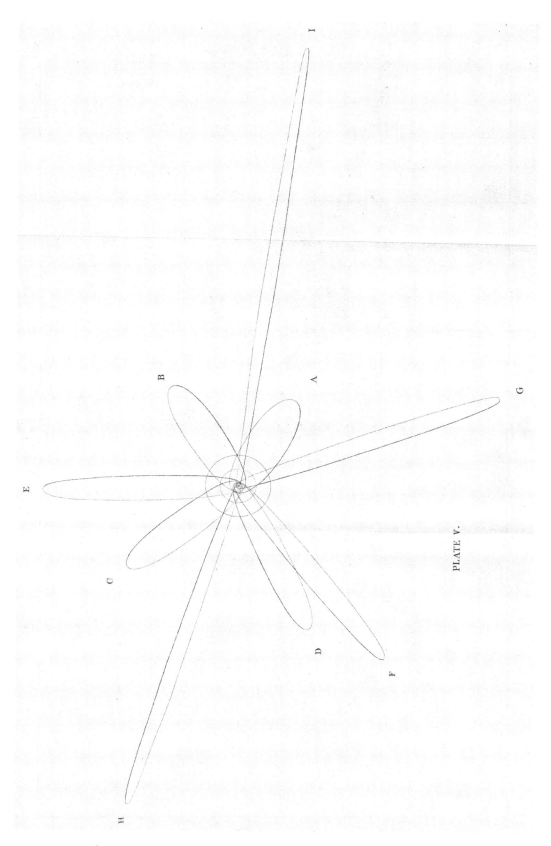

PLATE V.

The Scale of this Syftem is equal to one Third of the former.

Here I muft obferve to you, as a Thing I judge may prove of great Confequence with regard to the Syftem of Comets, which is as yet very imperfect: That I am ftrongly of Opinion, that the Comets in general, through all their refpective Orbits, defcribe one common Area, that is to fay, all their Orbits with regard to the Magnitude of their proper Planes, are mathematically equal to one another; which, if it once could be proved, and confirmed by Obfervation, the Theories of all the Comets that have been juftly obferved, might eafily be perfected, and their Periods at once determined, which now we can only guefs at, or may wait whole Ages for more Certainty of. What leads me to believe, that this may prove to be really the Cafe is this.

I find by Calculation, that the Orbits of the two laft Comets, whofe Elements have been moft corrected by Sir *Ifaac Newton* and Dr. *Hally*, are to one another, according to their Numbers, nearly as * 13 to † 17, notwithftanding one of them is one of the moft erratick that ever came under our Obfervation; and the other one of the moft neighbouring to the Sun.

But it is well known to all Mathematicians, that the firft of thefe Comets moved in fo eccentric a Trajectory, that the leaft Error in its almoft incredible Proximity to the Sun will produce a very fenfible Difference in the Area of the Orbit: And accordingly, if we moderate the Perihelion Diftance of this Comet, by making it but 1000 inftead of ‡ 612, which is but increafing it a $\frac{1}{33000}$th Part of the great Radius of the Orbit, (which is an Error every Aftronomer will readily grant is very eafily made) and we fhall find the Orbits of the faid two Comets to be exactly equal.

Further, I muft inform you, that the Comet of 1682, which the above compared with, feems to have been fo accurately obferved, that it does not appear to have altered its Perihelion Diftance half a 68th Part in one intire Revolution. Now, if we can with any Show of Reafon, and a Probability on our Side, bring the Areas of thefe two extream Comets, as I may call them, to an *Equality*, fure we may conclude, it is a Subject highly worthy to be more confidered and enquired into.

PLATE

---

* 1316539,968282 Comet of 1680.
† 1708155,4644 Comet of 1682.
‡ The Number in Dr. *Hally's* Synopfis.

PLATE VI.

Is a true Reprefentation of the fatellite Syftems, proportionable to one a-nother, and to the Orb of the Sun s Body, that a juft Idea of the Diftances of thofe fecondary Planets, may be eafier had from their refpective primary ones.

S reprefents the folar Body with its Atmofphere. *Fig.* 1. is the Syf-tem of *Saturn* from the fame Scale. *Fig.* 2. that of *Jupiter* from *ditto.* And *Fig.* 3. the Orbit of the Moon round the Earth, in the fame Proportion.

But as you can have but a very imperfect Idea of the Magnitude of thefe laft Circles, with regard to the Body of the Earth or Moon,

PLATE VII.

Is a true Projection of their real Globes, at their proper Diftance from each other, with their common Center of Gravity, and the Point and Line of equal Sufpenfion betwixt them, *viz.*

A, reprefents the Globe of the Earth.

B, that of the Moon.

C, Point, and C D, Line of equal Sufpenfion betwixt them.

E, Common Center of Gravity, which defcribes the *Orbis Magnus.*

E, F, and B, G, is the Orbit of the Moon.

Farther, that nothing may be wanting to give a true Notion of the whole together,

PLATE VIII.

Is a proportional Drawing of all the primary and fecondary Planets to-gether, diftinguifhed by their Characters, proper to attend a Globe of twelve Inches Diameter, fuch a one being fuppofed to reprefent the Sun.

PLATE IX.

Is an exact Scheme of the principal known Comets, in juft Proportion, to the Globe of the Earth reprefented at A, with the Nuclus, and Part of the Tail of the Comet of 1680, B, as it was obferved in its Affent from the Sun, *viz. a a* the Comet's natural Atmofphere, *z z z*, the *Denfer Matter* winding itfelf into the Axis of the Train *x x*, the inflam'd Atmofphere and Tail dilated near the Sun. C, reprefents the Ball of the Comet of 1682, D, that of 1665, E, that of 1742, and F, the Head of the Comet of 1744.

And again, that you may have fome Notion of the apparent Magnitudes of all thefe Planets and Comets, &c. as they appear at the Earth,

PLATE

PLATE VI.

Figure. I.

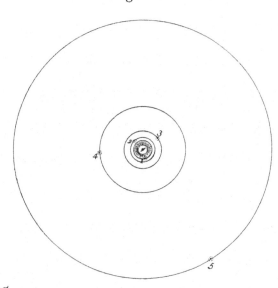

S

Fig. III

Fig. II

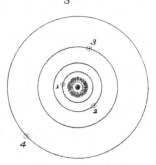

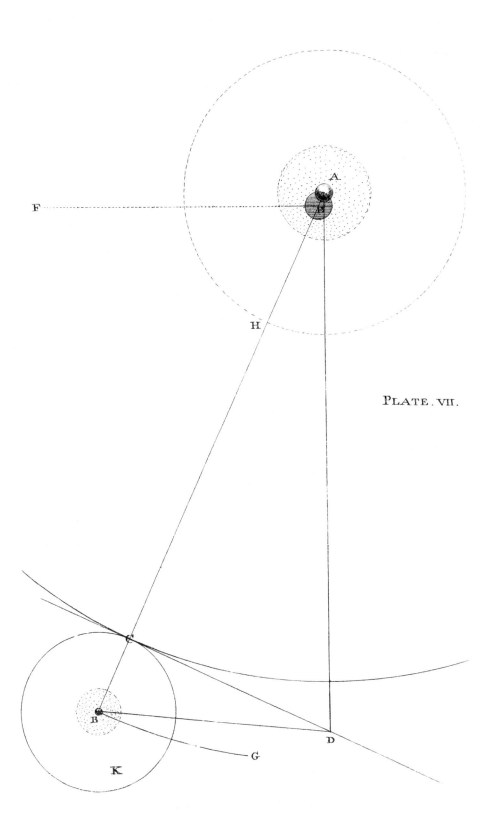

PLATE. VII.

PLATE VIII.

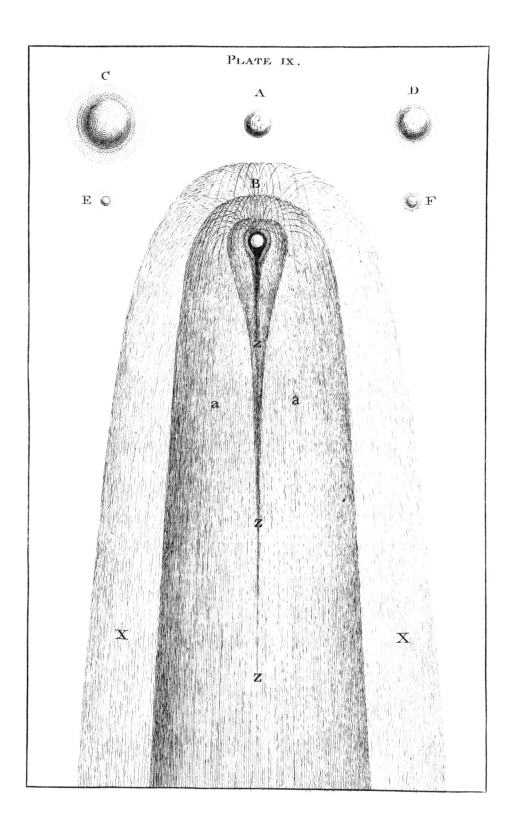

PLATE IX.

PLATE X.

PLATE X.

Reprefents the Sun and Moon in the juft Proportion of their mean Diameters, with two of the Comets A and B, and the five erratick Planets, as they are obferved at the Earth, in a middle State of their Diftances from it.

For a more full and particular Defcription of all the Parts of the folar Syftem, and of the home Elements of Aftronomy in general, I refer you to my *Clavis Cæleftis*, &c. where every thing concerning the Planets, Comets, and Stars ; and their real and apparent Motions, are at large reprefented, explained, and accounted for, for the Benefit of fuch as have not made the Mathematicks their regular Study.

Now, to convince you that the Planets are all in their own Nature no other than dark opaque Bodies, reflecting only the borrowed Light of the Sun, I muft recommend to your Obfervation, this natural and fimple Experiment, which almoft any Opportunity of feeing the *Moon* a little before the Full, will put into your Power to make ; but beft and eafieft when the Sun is in any of the North Signs, *i. e.* in *Summer.*

At fuch a time, the Sun being near fetting, the Moon will appear in the eaftern Hemifphere ; and if there be any bright Clouds northward, or fouthward near her, you will plainly perceive, that the *Light* of the one is of the fame Nature with that of the other ; I mean the Light of the Moon, and that of the Cloud. To me there never appeared any Difference at all ; and I am perfwaded, were you to make but two or three Obfervations of this kind, which is from Nature itfelf, a fort of ocular Demonftration, you cannot fail of being convinced, that the Moon's Light, fuch as it is, without Heat, can poffibly proceed from no other Caufe than that which illumines the Cloud : For if the Clouds, whofe Compofition we know to be but a thin light Fluid, formed of condenfed Vapours only, is capable of remitting fo great a Luftre, how much more may we not allow the Moon, which, Length of Time, and many other Circumftances, have long confirmed to be a durable and folid Body.

The Increafe of her Luftre, indeed, during the Abfence of the Sun from us, to a lefs penetrating Genius than your's, may poffibly afford fome trifling Ground of Objection to the above Conclufions, as being drawn from the Phænomena of Day-light only ; by reafon in the Night, we have no Clouds in equal Circumftances to compare with her.

But this I need not tell you, is all owing to her being feen through a darker Medium, and not to any real Increafe of natural Light emitted from the Sun. As a Proof of which, were it neceffary, you need only, fhut out the Rays of the Atmofphere, by the Help of a fufficiently

long

long Tube; and the Moon, or any other celeſtial Body, will appear through it, as bright in the Day-time as in the Night.

Thus all light Bodies of inferior Luſtre, whether ſhining by their own natural Radiences, or by a borrowed Reflection, partake of the ſame Advantage, when removed from the more potent Influence of a ſuperior one; and hence it is, that the *Aura Ætherea ſhines out moſt manifeſt, when the Body of the Sun himſelf is hid, the Stars, and the Via Lactea moſt lively and numerous in the Abſence of the Moon, and thoſe Exhalations, or Meteors, vulgarly called Falling-ſtars, become only viſible (like Glow-worms) in the Night.

Here it may not be improper to tell you, that the Clouds are to us in effect no other than as ſo many Moons, whereby we have our artificial Day prolonged to us ſeveral Hours after the Sun is ſet, and likewiſe produced as much ſooner before he riſes; and were they to aſcend by ſtill ſtronger Power of Exhalation to an Elevation, all round the Atmoſphere, ſo as to form a Sphere equal to four Times the Globe of the Earth, there would then be no ſuch Thing as real nocturnal Darkneſs to any Part of the World.

The lunar Light then we may very juſtly conclude, proceeds originally from the Sun: And notwithſtanding many more Arguments might be drawn from the Demonſtration of her Phaſes, Eclipſes, &c. to prove it, yet none of them need here be added, to what has been already ſaid, to convince you of the Truth of it. This being granted, let us now conſider what Effect this, or a like Quantity of borrowed Light, would have, when removed to a much greater Diſtance.

I may, I think, ſuppoſe, that you know ſo much of Opticks as to underſtand, that all viſible Objects apparently decreaſe in Magnitude, as their Diſtance from the Eye increaſes. Conſequently, that, if the Moon's Orbit was placed as far again from the Earth as it really is, her Globe, or rather Diſk, would then ſeem to be but half as big as to us ſhe now appears to be, and of courſe ſtill farther, were ſhe placed at ten times the Diſtance ſhe is known to revolve at, her apparent Diameter would be reduced to a tenth Part only of what it now appears to be in her preſent Orbit, that is, one hundred Times leſs in viſible Magnitude than her neighbouring Diſk is found to be where it now is ſeen. And ſuch, but ſomething leſs, the two Planets Venus and Jupiter, which are frequently, in their Turns, our Morning and Evening Stars, appear to be through a common Teleſcope.

Now

---

* An Helios, or golden Light, always attending the Sun, and ſuppoſed to ſpread itſelf all round his Body in the Direction of his Equator, was very viſible during the total Darkneſs of the Eclipſe of 1715, and may be always ſeen about the Autumnal Equinox.

Now thefe two Planets, together with the other three, which we find moving in regular Orbits round the Sun, are all found fubject to the fame * Changes of *Phænomena*, in their various Afpects with the Sun; and who can doubt but that they are all of the fame or like Nature? But you'll fay, perhaps, how are we fure that *Venus* and *Jupiter* have no native Light of their own, fince many of the ancient Philofophers, and in particular *Anaximander*, allowed even the Moon to have fome; and befides, in Philofophy, as well as in Logick, I think you hold there is no proving a Negative, at leaft at fuch a Diftance.

To make you conceive the Impoffibility of fuch a Light, and next to a Demonftration, convince you of the Unnaturalnefs of fuch a Suppofition, I muft put you in mind, that fome time ago, when I was laft in the Country with you, I think it was about the latter End of Autumn, near the Winfter Solftice, as we were walking one Evening, I bid you take notice of the Moon, which was then near fetting, and about two Days old. You may remember, her whole Globe appeared to us very confpicuoufly within a manifeft Circle. You immediately told me, that that kind of Phænomenon the Country People called a *Stork*, or the old Moon in the new one's Arms. This I then endeavoured to explain to you, and I think made you fenfible it was intirely an Effect of the Earth's, and an Appearance always to be expected at that Time of the Year. The Earth being then in the State of a Full-Moon to that Part of the lunar Orbit, and near her Perihelion, at which time, the Earth fends back a Reflection to the † Moon twenty-five times more potent than that of the Moon to us.

Now the Planet *Venus*, from undeniable Principles of Geometry, is allowed to be nearly fuch another Globe as the Earth is; and fince the Earth, as I have juft now related, is found to reflect much more Light to the Moon by reafon of her fuperior Magnitude, than the Moon can poffibly reverberate to Earth again; and fince alfo 'tis plain, the Earth has no Light of its own, why then fhould we imagine *Venus* to be endowed with a Luftre, which we can prove to be no more than a fimilar Body, and governed by the fame Laws as the Earth is?

Anaximander's Miftake, in fuppofing the Moon in fome fmall Degree a radiant Body of itfelf, lay, in not confidering, that the faint Illumination here defcribed, and vifible all over her Globe, foon after almoft every Conjunction with the Sun; and probably in Eclipfes, alfo proceeded from the Earth; but the thing I think is too evident to expect any fort of Con-

E

tradiction,

---

* *Venus* and *Mercury* in every Heliocentrick Revolution, perform all the Changes of our Moon in a like Gradation and Defection of Light, both horned and gibos'd.
† Their Diameters being nearly as 1 to 5.

tradiction, therefore I hope you will admit it as a Truth, and confequently take it for granted, that the planetary Bodies in general, are meer terreftrial, if not terraqueous Bodies, fuch as this we live upon; which is the Thing I have chiefly in this Letter attempted to demonftrate, or have rather explained; and now I hope, for the future, you will receive the Idea of a Plurality of Worlds more favourably, and look upon aftronomical Conjectures in a lefs ridiculous Light than you ufed to do, efpecially fince you muft allow, they give our unlimited Imaginations a like all endlefs Field of Contemplation, not only full of the wonderful Works of Nature, but alfo of a vifible Providence.

I think I cannot conclude this Letter to you more properly, than with the following fine Lines of Mr. *Addifon*'s from the *Spectator*, Vol. VI. No. 465. which I hope you are not fo polite as to look upon as an unfafhionable Quotation.

> The fpacious Firmament on High,
> With all the blue ethereal Sky,
> And fpangl'd Heav'ns, a fhining Frame,
> Their great Original proclaim:
> Th' unwearied Sun, from Day to Day,
> Does his Creator's Pow'r difplay,
> And publifhes to ev'ry Land
> The Work of an Almighty Hand.
> Soon as the Ev'ning Shades prevail,
> The Moon takes up the wond'rous Tale,
> And nightly to the lift'ning Earth,
> Repeats the Story of her Birth:
> Whilft all the Stars that round her burn,
> And all the Planets in their Turn,
> Confirm the Tidings as they roll,
> And fpread the Truth from Pole to Pole.
> What though, in folemn Silence, all
> Move round the Dark terreftrial Ball?
> What tho' nor real Voice nor Sound
> Amid their radiant Orbs be found?
> In Reafon's Ear, they all rejoice,
> And utter forth a glorious Voice,
> For ever finging, as they fhine,
> " *The Hand that made us is divine.*"

And am, &c.

LETTER

LETTER the FOURTH.

Of the Nature of the heavenly Bodies continued, with the Opinions of the Antients concerning the Sun and Stars.

S I R,

YOU tell me you begin to be a tolerable good *Copernican*, and would now be glad to have my Opinion further upon the Nature of the Sun and Stars, with regard to the Suggeſtion of their being like Bodies of Fire. This you ſay will go a great Way towards confirming you in the Notion you have begun to embrace of a Plurality of Syſtems, and a much greater Multiplicity of Worlds than our little ſolar Syſtem can admit of. Beſides, ſhewing in a very evident Light, that the Authorities cited in my firſt Letter are founded upon the cleareſt Reaſon.

Anaxagoras, you ſay, believed the Sun to be a Lump of red-hot Iron; *Euripides* thought it a Clod of Gold; and others ſtill more ridiculouſly have imagined it to be a dark Body, void of all Heat. That the Sun is a vaſt Body of blazing Matter, notwithſtanding the various Opinions of thoſe primitive Sages, will, I think, hardly admit of a Queſtion: Since the known Warmth of his prolifick Beams, and the viſible Effect of the Burning-glaſs, puts it quite out of the Power of our preſent Set of Senſes, at leaſt to argue againſt it; and how reaſonably we may imagine the Stars to be all of the ſame or like Nature, will ſufficiently appear from theſe following Conſiderations: Firſt, it is well known to all Mathematicians, that any viſible Object of any determined Magnitude may be reduced to the Appearance of * a phyſical Point, by removing the Eye of the Obſerver to a proper or proportionable Diſtance from it, within the finite View: And that the apparent Diameter of every luminous celeſtial Body, will always be diminiſhed reciprocally, in Proportion to the Diſtance from the Eye, till they become altogether imperceptible.

E 2   Thus

---

* What is here meant by a phyſical Point, is a Point viſible to the naked Eye, which human Art cannot divide; and ſo far it partakes of the Property of a mathematical one, which is only to be conceived, and not ſeen.

Thus the Difk of the Sun, which appears to us at Earth under an Angle of about half a Degree, if feen from the Planet *Saturn*, would appear not much bigger than the Planet *Venus* or *Jupiter*, in their moft neighbouring Vicinity does to us; and confequently to an Eye placed in the Aphelion Point of the Orbit of the great *Comet* of 1680, his apparent Diameter would be fo reduced as to feem but little bigger than the largeft of the Stars; and by the fame Analogy, or Way of Reafoning, admitting Space and Diftance infinite, which I humbly apprehend is not to be difputed, were all the Matter in the Univerfe united, and conglobed in one Mafs, with refpect to ocular Senfation, it might be diminifhed fo near to a ma- thematical Punctum, as to be almoft adequate to our Ideas of Nothing.

This to any tolerable Optician, muft be an evident Conviction of the Truth of the modern Aftronomy, which now univerfally allow all thofe radiant Bodies the Stars to be of the fame Nature with the Sun; and that as certainly they are no other than vaft Globes of blazing Matter, all un- doubtedly fhining by their own native Light.

But as you have often objected to what has been faid of the Diftance of the Stars in general, and may poffibly from a Suppofition, that they are, or may be, much nearer to us, infer, that their Light, like that of the Planets, may be alfo borrowed from the Sun, or from fome other radiant Body, which, from the Nature of the Suppofition, muft of Confequence be in- vifible to us, I judge it will not be amifs to throw a few demonftrative Ar- guments in your Way, in order to lead you a little out of the Path of an early Prejudice, and draw you as it were by Degrees through the Dawn of aftronomical Reafoning, out of your original Error, and refcue your Imagination from the falfe Notions imbibed from Phænomena only in your younger Years. This I guefs cannot fail of reconciling you to this more rational Way of Thinking, and make you acquainted with Truths of much Confequence, which perhaps you have yet been an intire Stranger to. The grand *Deceptio Vifus*, which I muft firft endeavour to remove, and which as a fort of Paradox in Nature, has, as I may fay, im- prifoned the Underftanding of many fuperficial Reafoners, and in general all incurious Men, is this.

Moft People are too apt to think originally, that as the Heavens appear to be a vaft concave Hemifphere, that the Stars muft of courfe, as of Confequence, be fixed there, like fo many radiant Studs of Fire, of various Magnitudes; and take it for granted, chiefly defigned for no other Purpofe than to deck and adorn the Canopy of our Night. This was long ago the Opinion of *Thales* the *Milefian*, and wants not the Authority of

many

many of the Antients to back it. Others, in particular * *Ptolomy* of *Pelusium* in *Africa*, who from his Experience in this Science, is called by some the Prince of Astronomers, believed them to be Loop-holes in the vast solid celestial Firmament emitting the Light of the Crystalline Heaven through it to all within it. The famous *Diogenes*, Cotemporary with *Plato*, conceived them to be of the Nature of Pumice-stones, and inclined to an Opinion, that they were the *Spiracula*, or Breathing-holes of Heaven. *Anaxagoras* thought them Stones snatched up from the Earth by the Rapidity of its Motion, and set on Fire in the upper Regions above the Moon.

But how ridiculous and absurd all these Opinions and Conjectures really are, will easily appear, if we but once consider the Nature of an unbounded Æther, and the amazing Property of infinite Space.

This, with what has been said before, will not a little assist your Imagination towards conceiving the Reasonableness of the Notion modern Astronomers are now confirmed in, of their being absolutely so many burning Balls, and which was no doubt, many Years ago, the Opinion of *Manilius*, as is evident from these Lines in his Poem of the Sphere.

> For how can we the rising Stars conceive
> A casual Production ; or believe
> Of the chang'd Heav'ns the oft renascent State
> *Sol*'s † frequent Births, and his quotidian Fate.
>
> SHERBURNE.

And again in the same Poem:

> The fiery Stars, and Æther that creates
> Infinite Orbs, and others dissipates.
>
> *Zoroaster,*

---

* *Ptolomy* supposed two Heavens above that of the fixed Stars, which he called the eighth ; *viz.* a ninth, the Crystalline, and a tenth the *Primum Mobile.* See Letter the second.

> The sacred Sun, above the Waters rais'd,
> Thro' Heav'ns eternal, brazen Portals blaz'd ;
> And wide o'er Earth diffus'd his chearing Ray,
> To Gods and Men to give the golden Day.
>
> HOMER.

† *Xenophanes* believed the Stars to be no other than Clods set on Fire, quenched in the Daytime, and rekindled in the Night.

Zoroafter, the firſt of all Philoſophers we read of who ſtudied the Stars, is reported to have believed them of a fiery Nature. *Empedocles* judged them to be Fire æthereal, ſtruck forth in its Secretion, and blazing in the upper Regions. *Plato* thought them Fire, with the Mixture of other Elements as Cements. *Heraclides* Worlds by themſelves, of *Earth, Air,* and *Fire* ; and *Ariſtotle*, ſimple Bodies of the Subſtance of Heaven, but more condenſed.

But that I may not take up too much of your Time with Opinions that has been imbibed in the Infancy of Aſtronomy, and has long ago been exploded, I ſhall attempt but one Thing more to confirm your Sentiments in this new Doctrine.

Firſt, that the Stars are all at a Diſtance, not to be determined by the utmoſt Perfection of human Art, is manifeſt from their having very little, or no ſenſible Parallax ; and conſequently, that any one of them is abſolutely bigger or leſs than another, from the ſimple Laws of Opticks, cannot poſſibly come under our Obſervation to be aſcertained ; but that they all of them may be nearly of the ſame Size or Solidity, is as impoſſible, with any Shew of Reaſon to deny, ſince it is a known Principle in Geometry, that all viſible Objects naturally diminiſh, as has been ſaid before, or are magnified in a certain Proportion to their Diſtance from the Eye ; and hence we may conclude, and not without Reaſon in its ſtrongeſt Light to ſupport us, that the ſmalleſt Stars, to the very leaſt Denomination, are only removed reſpectively more diſtant from the Obſerver's Station ; and that at leaſt this we may be certain of, that they are all together undoubtedly an Infinity of like Bodies, diſtributed either promiſcuouſly, or in ſome regular Order throughout the mundane Space: And, as *Marino* ſays,

> Reſplendent Sparks of the firſt Fire !
> In which the Beauty we admire,
> And Light of thoſe eternal Rays,
> The uncreated Mind diſplays.

It remains now I think to ſhew, and endeavour to prove, that the Stars are not only light Bodies of the Nature of the Sun, but that they are really ſo many Suns, all performing like Offices of Heat and Gravity, in a regular Order, throughout the viſible Creation, in oppoſition to an Opinion

you

---

* Mr. *Bradley*, Aſtroromer-Royal, has, in a great meaſure, proved that the Aberration of the Stars hitherto miſtaken for a Parallax, may ariſe from, and indeed ſeems to be no other than the progreſſive Motion of Light, and Change of Place to the Eye, ariſing from the Earth's annual Motion and Direction.

you have formerly hinted at, of their being in another Senfe of a fecondary Nature.

All Objects within the fenfible Sphere of the Sun's Attraction, or Activity, are in fome meafure magnified by a good Telefcope: But the Stars are all placed fo far without it, that the beft Glaffes has no other Effect upon them than making them appear more vivid or lively, but all inate opaque Bodies, reflecting only a borrowed Light from fome primary one, contrary to this Property, are all obferved to lofe their Light, in the fame Proportion, as they are magnified, and through all Glaffes become more dull than otherwife they appear to the naked Eye: And hence we may infer, without any further Evidence, that the Stars are all light Bodies endowed with native Luftre; and that Bodies, like the known Planets, from the fame Reafoning, it is as clear they cannot be, becaufe their Diftance, though uncertain as to the Truth of the whole, yet fuch a Part of it as cannot be denied, would render them all in fuch a Cafe invifible.

A Proof of this will plainly prefent itfelf, if we confider the Courfe of the known Comets, who all of them, without Exception, become imperceptible, and intirely difappear; though moft of them much bigger than the Earth, or any of the leffer Planets, long before they arrive at their refpective Aphelions.

But we are under a kind of Neceffity to believe them either Suns or Planets, that is either dark or light Bodies; and fince I have fhewn the Improbability; nay, I may venture to fay, the Impoffibility of their being the firft, it is natural fure to conclude, that they muft be of the laft Sort; and I am perfuaded, if you but once confider how ridiculous it is to imagine fo vaft a Number of Bodies, all rolling round a Number of invifible Suns, which muft otherwife be the Cafe, fince they are feen on all Sides of ours, and cannot poffibly be enlightened by him, or any, how all of them, by any one elfe, you cannot poffibly have any fort of Difficulty in this Determination: But that no Arguments may be wanting to enforce your Belief of what is here concluded, it will not be amifs to put you in Mind of an optical Experiment or two, which cannot fail of convincing you of the vaft Probability of what is here afferted of them; and next to a moral Certainty, demonftrate the Truth of what fo many of the beft Aftronomers have advanced, as before namely, that the Stars are all, or moft of them, Suns like ours.

Place any concave Lenfe before your Eye, and you will find all vifible Objects will appear through it, as removed to a much greater Diftance than they really are at, and reciprocally as much diminifhed. Now, if
you

you look upon one of thefe Glaffes of a proper Concavity, oppofed to the Sun or Moon, you will refpectively have the Appearance of a real Star or Planet, the firft exhibited by the Body of the Sun, the other by the Moon, and either more or lefs diminifhed in Proportion to the Surface of the Sphere the Glafs is ground to.

For Example, a double Concave, or Glafs of a negative Focus, ground to a Sphere of about three Inches Diameter, will if oppofed to the Sun's Difk at a proper Diftance from the Eye, help you to a very good Idea how the Sun appears to the Planet *Jupiter* ; and if a proper Regard be had to the Diftance of the Planet *Saturn*, a Lenfe ftill more concave may be formed to give a juft Idea of the Sun's Appearance to *Saturn*. Again, one much more concave than the former, proportioned to the Orbit of *Mars,* will naturally exhibit the folar Body, as feen from that Planet.

To the Planet *Venus* and *Mercury*, the Sun appearing much larger than to us at the Earth, to have any tolerable Notion of his varied Phænomena to them, it will be neceffary to procure Glaffes of a fuitable Convexity, ground to reciprocal Concaves, which may eafily be done to any Focus, fo as to fhew how the Sun, naturally appears to the Inhabitants of thofe two Planets.

The various Appearances of the Planets themfelves to us at the Earth, may alfo well enough be had, if through Glaffes analagous to their refpective Diftance and Magnitude, we look at the Moon, particularly all the Phafes of *Venus*, and even of *Mercury*, and the Gibofity of *Mars*, *&c.* may be juftly and beautifully reprefented at different Ages of the Moon, as thofe Planets appear through the largeft and beft Telefcopes.

This Way you may convince even your Friend * * *, who you tell me has reafoned all his Senfes ufelefs, and yet continues fo great an Atheift in Aftronomy, as not to believe the World turns round upon its Axis, though he gives no better Reafon for it than that of his not being giddy.

After all thefe Arguments, I hope no new Difficulties will arife to retard your Belief, or deprive the Stars of their folar Nature, fo juftly due to them : This Point gained, the next Thing to be confidered is, whether all thofe glorious Bodies, the far greater Part of whom being invifible to the naked Eye, were made purely and purpofely for the fole Ufe of this diminitive World, our little trifling Earth.

> ————Men, conceited Lords of all,
> Walk proudly o'er this pendent Ball,
> Fond of their little Spot below,
> Nor greater Beings care to know,
> *But think thofe Words, which deck the Skies,*
> *Were only form'd to pleafe their Eyes.*   DUCK.

The

The very Suppofition not only implies a profound Ignorance of the Divine Attributes, but is as impious, and full of Vanity, as it is erroneous and abfurd, and even a Blindnefs fufficient of itfelf, were there no other Caufe for it, to introduce Idolatry in the Minds of Mortals, by finking the divine Nature fo near to the human.

It being granted that the Stars are all of the fame Kind, I think it may be agreed, that what we evince of any one may be allowed to be true of any other, and confequently of all the reft. This *Poftulata* gained, I fhall next proceed to enquire what the real Ufe and Defign of fo many radiant Bodies are, or may be made for.

The Sun we have juftly reduced to the State of a Star, why then in Reafon fhould he have his attendant Planets round him, more than any of the reft, his undoubted Equals? No Shadow even of a Reafon can be given for fuch an Abfurdity.

May we not with the greateft Confidence imagine, that Nature as juftly abhors a *Vacuum* in Place, as much as Virtue does in Time? Surely yes: And by fuppofing the Infinity of Stars, all centers to as many Syftems of innumerable Worlds, all alike unknown to us; how naturally do we open to ourfelves a vaft Field of Probation, and an endlefs Scene of Hope to ground our Expectation of an *ever*-future Happinefs upon, fuitable to the native Dignity of the awful Mind, which made and comprehends it; and whofe Works are all as the Bufinefs of an Eternity?

If the Stars were ordained merely for the Ufe of us, why fo much Extravagance and Oftentation in their Number, Nature, and Make? For a much lefs Quantity, and fmaller Bodies, placed nearer to us, would every Way anfwer the vain End we put them to; and befides, in all Things elfe, Nature is moft frugal, and takes the neareft Way, through all her Works, to operate and effect the Will of God. It fcarce can be reckoned more irrational, to fuppofe Animals with Eyes, deftined to live in eternal Darknefs, or without Eyes to live in perpetual Day, than to imagine Space illuminated, where there is nothing to be acted upon, or brought to Light; therefore we may juftly fuppofe, that fo many radiant Bodies were not created barely to enlighten an infinite Void, but to make their much more numerous Attendants vifible; and inftead of difcovering a vaft unbounded defolate Negation of Beings, difplay an infinite fhapelefs Univerfe, crowded with Myriads of glorious Worlds, all varioufly revolving round them; and which form an Atom, to an indefinite Creation, with an inconceivable Variety of Beings and States, animate and fill the endlefs Orb of Immenfity.

F   That

That the fidereal Planets are not vifible to us, can be no Objection to their actual Exiftence, and being there, is plain from this; it is well known, that the Stars themfelves, which are their Centeral, and only radiant Bodies, are little more to us at the Earth, than mathematical Points. How ridiculous then is it to expect, that any of their fmall opaque Attendance, fhould ever be perceived fo far as the Earth by us; and befides, to fhow the Impoffibility of fuch a Difcovery, we need only confider, what is, and what is not to be expected, or known in our own home Syftem. All the Planets in this our fenfible Region, every Aftronomer knows, is far from being vifible to one another, in every individual Sphere; for to an Eye at the Orb of *Saturn*, this Earth we live upon, which requires Years to circumfcribe, and Ages to be made acquainted with, and is far from being yet all known, cannot poffibly from the above Planet be feen: And further, fince *Saturn* and *Jupiter*, two of the moft material and confiderable Globes we know of, except the Sun himfelf, are Bodies apparently of the fame kind, and are obferved to have each a Number of lefler Planets moving round them; why may we not expect with equal Certainty and Propriety, that all other Bodies, under the fame Circumftances, are in like manner attended; that is, feeing the Sun is found to be the Center of a Syftem of Bodies, all varioufly volving round him? where lies the Improbability of his fellow Luminaries, the Stars, being furrounded in like fort, with more or lefs of fuch Attendance.

I fhall offer but one Thing more to your Confideration in this Affair, and which I am in great Hopes will be fufficient to make you think thefe natural Suggeftions a good deal more than probable, and that is this:

The modern Aftronomers having, in a great meafure, proved that the Stars are, in all refpects, vaft Globes of Fire like our Sun. Let us fuppofe a new-created Mind, or thinking Being, in a profound State of Ignorance, with regard to the Nature of all external Objects, but fully endowed with every human Senfe and Force of Reafon, fufpended in Æther, exactly in the midway, betwixt * *Syrius* and the Sun; in which Cafe, both of thefe Luminaries would equally appear much about the Brightnefs of the largeft of our Planets. Now fhould fuch a Being, determined either by Accident or Choice, arrive at this our Syftem of the Sun, and feeing all the planetary Bodies moving round him, I would afk you what you think he would imagine to be round *Syrius?* Your Anfwer, I think I may venture to fay, would not be *nothing*; and methinks I already hear you fay, Why Planets fuch as ours.

---

* A Star of the firft Magnitude in the greater *Dog*, and the moft neighbouring to our Sun.

PLATE

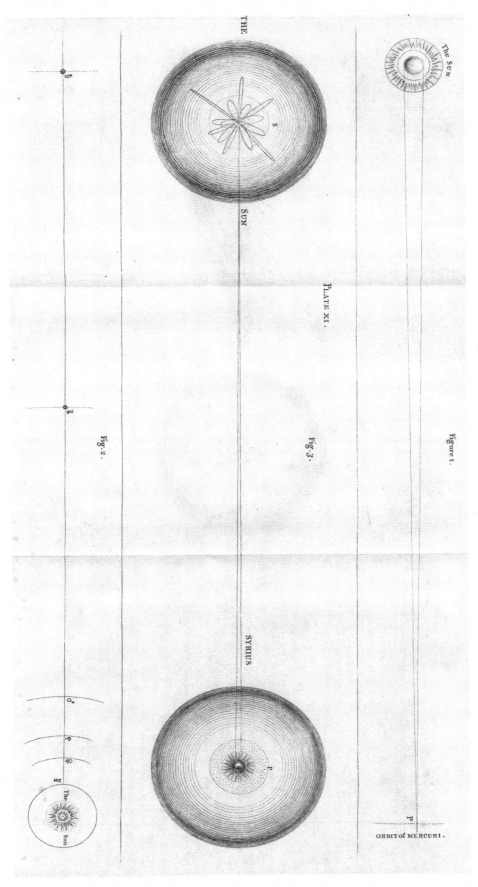

PLATE XI.

Is defigned as a geometrical Scale to all the primary Parts of the vifible Creation, with regard to the Diftance of Orbits compared with the Globe of the Sun; by which at once may be conceived, and juftly meafured in the Mind, not only the mean Diftance of the Planets with regard to one another, but alfo that of the Comets, and even the comparative Diftances of the neareft of the Stars, which will, I guefs, greatly help you to form an Idea of the vaft Extent of Space neceffary to comprehend the whole Creation.

Fig. 1. Is a Radius of the Orbit of *Mercury*, in true Proportion to the Body of the Sun reprefented at S, fhewing at the fame time a fmall Portion of the opaque Planet's Orbit, and the real Length of its Shadow at P.

Fig. 2. Is a Radius of the whole Syftem of the Planets as far as the Orbit of *Saturn* in Proportion to a compleat Orbit of *Mercury*, much lefs than the former; the former ferving as a better known Scale to confider the amazing Diftances of the more remote Planets by.

Laftly, *Fig.* 3. Is a Reprefentation of the leaft poffible Diftance of *Syrius* and the Sun, proportionable to the Magnitude of the Sphere of our Comets, &c. reprefented at S, whereby it evidently appears, that as all the Planets of *Syrius* muft be included within the fmall Sphere reprefented in the Center P, none of them could poffibly be feen at the Sun, not only by reafon of the Smallnefs of the Angle of Suftenfion, or Elongation, but alfo as being loft in the fuperior Light of *Syrius* himfelf, in fo minute an Orb of Vicinity.

Confequently (as you muft perceive) no Arguments can poffibly be drawn to deny the Exiftence of fuch Bodies, with any Shew of Reafon, from their not having been feen by us.

Here I muft obferve to you, that you cannot confider this Scale of Orbits too much before you look upon Plate XVII.

To conclude, it evidently feems to be the End and Defign of Providence, by this vifible Variety of Beings, to lift the Minds of Men above this narrow Earth, in Search of that powerful Being upon which we are all fo much dependant; and the *Creator*, no doubt, in this vaft Difplay of his Wifdom and Power, defigned the amazing Whole, as the adequate Object of every Part, and as fuch equally open on all Sides, to the penetrating Progrefs of human Minds, and through the moft extenfive Faculty of Senfe, the *Sight*, to draw our Reafon and Underftanding by Degrees, from finite Objects into Infinity; and as the laft Refult of celeftial Contemplations place within our Reach, a certain Evidence of a future State, *and the manifeft Manfions of Rewards and Punifhments, fuited no doubt moft equitably to all Degrees of Virtue, and to every Vice.*

When

" When I confider (fays Mr. *Addifon*, fpeaking as having taken particular
" notice of a fine Evening) that infinite Hoft of Stars, or to fpeak more
" philofophically of Suns, which were then fhining upon me, with thofe
" innumerable Sets of Planets or Worlds, which were then moving round
" their refpective Suns; when I ftill enlarge the Idea, and fuppofed ano-
" ther Heaven of Suns and Worlds rifing ftill above this which we dif-
" covered; and thefe ftill enlightened by a fuperior Firmament of Lu-
" minaries, which are planted at fo great a Diftance, that they may ap-
" pear to the Inhabitants of the former as the Stars do to us; in fhort,
" whilft I purfued this Thought, I could not but reflect on that little
" infignificant Figure which I myfelf bore amongft the Immenfity of
" God's Works:" This Reflection, I judge, as you are an Admirer of the
Author, you will not look upon as impertinent in this Place, efpecially as
it muft enforce what I have endeavoured to fhew you, namely, the Rea-
fonablenefs of a Plurality of fidereal Syftems, and their Multiplicity of
Worlds; which, if you are yet in Doubt of, I hope you will at leaft for-
give fo well defigned an Attempt with your ufual Candour.

I am now prepared to proceed in the chief Defign of this Undertaking,
which is to folve the Phænomena of the *Via Lactea*; and propofe in my
next to anfwer more fully your farther Requeft.

I am, &c.

LETTER

LETTER the FIFTH.

Of the Order, Distance, and Multiplicity of the Stars, the Via Lactea, *and Extent of the visible Creation.*

S I R,

WE are told, and, if I remember right, it is also your Opinion, that three of the finest Sights in Nature, are a rising Sun at Sea, a verdant Landskip with a Rainbow, and a clear Star-light Evening : All of which I have myself often observed with vast Delight and Pleasure. The first I have frequently beheld, and always with an agreeable Surprize ; the second I have as often taken notice of, with no small Degree of Admiration ; but the last I shall never look up to without an Astonishment, even mixed with a kind of Rapture. The Night you last left us, this admirable Scene was in its full Beauty ; and, as *Milton* says,

> Silence was pleas'd : now glow'd the Firmament
> With living Saphirs ; *Hesperus* that led
> The starry Host rode brightest.————

I found it was impossible to look long upon this stupendious Scene, so full of amazing Objects, and particularly the *Via Lactea*, which (the Moon being absent) was then in great Perfection, without being put in Mind of my Task. This surprizing Zone of Light being the chief Object I have undertaken to treat of and demonstrate.

This amazing Phænomenon which have been the Occasion of so many *Fables*, idle Romances, and ridiculous Opinions amongst the Antients, still continues to be unaccounted for, and even in an Age vain enough to boast Astronomy in its utmost Perfection.

What will you say, if I tell you, it is my Belief we are so far from the real Summit of the Science, that we scarce yet know the Rudiments of what may be expected from it. This luminous Circle has often engrossed my Thoughts, and of late has taken up all my idle Hours ; and I am now in

great

great Hopes I have not only at laſt found out the real Cauſe of it, but alſo by the ſame Hypotheſis, which ſolves this Appearance, ſhall be able to de-monſtrate a much more rational Theory of the Creation than hitherto has been any where advanced, and at the ſame Time give you an intire new Idea of the Univerſe, or infinite Syſtem of Things. This moſt ſurprizing Zone of Light, which have employed ſucceſſively for many Ages paſt, the wiſeſt Heads amongſt the Antients, to no other Purpoſe than barely to deſcribe it; we find to be a perfect Circle, and nearly biſecting the ce-leſtial Sphere, but very irregular in Breadth and Brightneſs, and in many Places divided into double Streams.

* The principal Part of it runs through the *Eagle*, the *Swan*, *Caſſiopea*, *Perſeus*, and *Auriga*, and continues its Courſe by the Head of *Monoceros*, along by the greater *Dog* through the Ship, and underneath the *Centaur's Feet*, till having paſſed the *Alter*, the *Scorpion's Tail*, and the Bow of *Aquarius*, it ends at laſt where it begun.

P L A T E   XII, and XIII.

Repreſents the two Hemiſpheres, where its true Tract is diſtinguiſhed amongſt the principal Stars, and may eaſily be conceived by them to cir-cumſcribe and biſect the whole Heavens.

This is that Phænomena I am about to explain and account for; but before I proceed farther, I judge it will be no *improper Precognita*, to give you the Thoughts of the Antients upon it; the Relation perhaps may re-quire ſome Patience; but I gueſs, that after reading ſuch wild and extra-vagant Notions concerning it, you will naturally judge more favourably of the Conjectures of the Moderns upon it, and particularly of what is con-cluded in the ſucceeding Pages.

Theophraſtus

* ————— Carried toward the oppoſed *Bears*,
Its Courſe cloſe by the *Artick* Circle ſteers,
And by inverted *Caſſiopea* tends;
Thence by the *Swan* obliquely it deſcends
The Summer Tropick, and *Jove's* Bird divides;
Then croſs the Equator, and the Zodiack glides
Twixt *Scorpio's* burning Tail, and the left Part
Of *Sagitarius*, near the fiery Dart;
Then by the other *Centaur's* Legs and Feet,
Winding remounts the Skies (again to meet)
By *Argos'* Topſail, and Heav'ns middle Sphere,
Paſſing the *Twins*, t' o'ertake the Charioteer;
Thence *Caſſiopea* ſeeking thee does run,
O're *Perſeus* Head, and Ends where it begun.

SHER. MANILIUS.

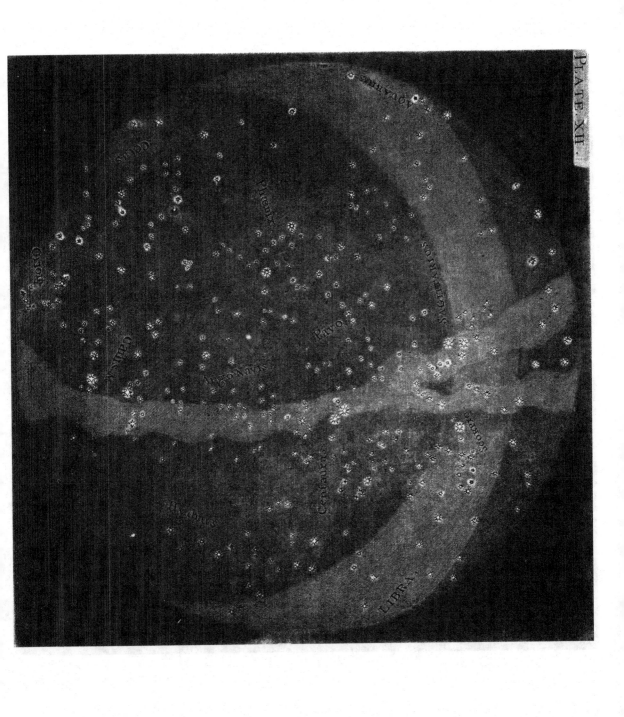

PLATE XII.

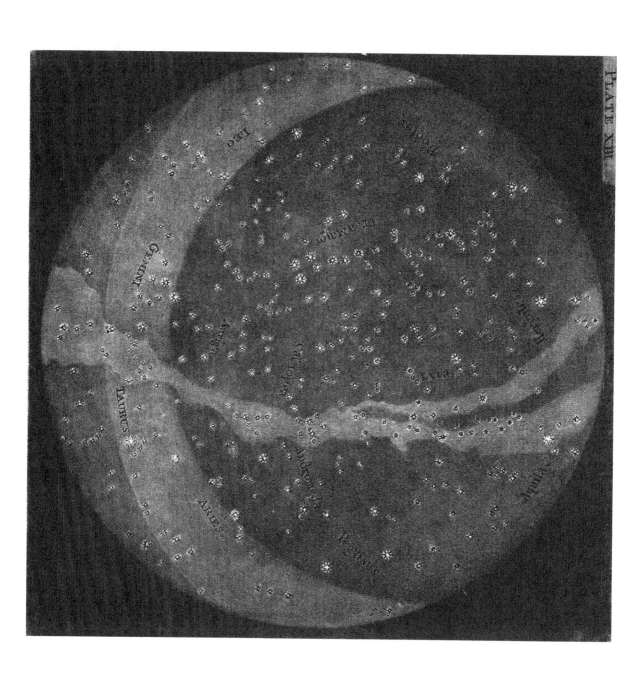

PLATE XIII.

LEO

Boötes

GEMINI

Ursa Major

Hercules

Auriga

Cassiopeia

Lyra

TAURUS

Andromeda

Aquila

ARIES

Pegasus

Theophraſtus * was of Opinion, that the Hemiſpheres, which, by many of the Antients were imagined to be ſolid, was joined together here ; and that this was the foldering of the two Parts into one. † DIODORUS thought it celeſtial Fire, of a denſe and compact Nature, ſeen through the Clifts or Cracks of the parting Hemiſphere : But as *Manilius* ſays,

> Aſtoniſhment muſt ſure their Senſes reach,
> To ſee the World's wide Wound, and Heav'n's eternal Breach.

OENOPIDES ‖ believed it the ancient Way of the Sun, till frighted at the bloody Banquet of *Thyeſtis*. ** ERATOSTHENES ſuppoſed it *Juno's* Milk, ſpilt whilſt giving Suck to *Hercules*. ‡‡ PLUTARCH makes it the Effect of *Phaeton's* confuſed Erratication ; but I think it is plain †† OVID judged them to be Stars, and the ancient *Ethnicks* believed them to be the blisful Seats of valiant and heroic Souls.

> ——Valiant Souls, freed from corporeal Cives,
> Thither repair, and lead æthereal Lives.
>   MANILIUS.

* *Macrobius*, lib. i. cap. 15.
> Or meets Heaven here ! and this white Cloud appears
> The Cement of the cloſe-wedg'd Hemiſpheres !

† The ſacred Cauſes human Breaſts enquire,
> Whether the heavenly Segments there retire,
> (The whole Maſs ſhrinking, and the parting Fame
> Thro' cleaving Chinks admits the ſtranger Flame.

‖ Or ſeems that old Opinion of more Sway,
> That the Sun's Horſes here once run aſtray,
> And a new Path mark'd in their ſtraggling Flight,
> Of ſcorching Skies, and Stars aduſted Light.

** Nor muſt that gentle Rumour be ſuppreſt,
> How Milk once flowing from fair *Juno's* Breaſt
> Stain'd the celeſtial Pavement, from whence came
> This milky Path, its Cauſe ſhewn in its Name.

‡‡ When from the hurried Chariot Light'ning fled,
> And ſcatter'd blazes all the Skies o'erſpread ;
> By whoſe Approach new Stars enkindled were,
> Which ſtill as Marks of that ſad Chance appear.
>   MANILIUS.

†† A Way there is in Heaven's expanded Plain
> Which when the Skies are clear, is ſeen below,
> And Mortals by the Name of *Milky*, know,
> The Ground-work is of Stars ----
>   *Ovid's* Met. lib. i.

                                   But

But * Democritus long ago believed them to be an infinite Number of small Stars; and such of late Years they have been discovered to be, first by *Gallaleo*, next by *Keplar*, and now confirmed by all modern Astronomers, who have ever had an Opportunity of seeing them through a good Telescope.

PLATE XIV.

Is from an Observation I made myself, of a bright Part of this Zone near the Feet of *Antinous* ; which, (by a Mistake of the Engraver) is, as it appears through a Tube of two convex Glasses.   I saw it through a very good Reflector, and formed the Plan by a Combination of Triangles.

Milton takes notice of this Zone in a most beautiful Manner, where he describes the Creator's Return from his six Day's Work to Heaven, he introduces it as a Simile to express his Idea of the eternal Way, or Road to the celestial Mansions.

——— A broad and ample Road, whose Dust is Gold
And Pavement Stars, as Stars to thee appear,
Seen in the *Galaxie*, that Milky Way,
Which nightly as a circling Zone thou seest
Powder'd with Stars.

But to infer from their Appearance only, that they are really Stars, without considering their Nature and Distance; and that nothing but Stars could possibly produce such an Effect, may perhaps be assuming too much, when we have nothing but the bare Credit of the *Belgic* Glasses to support our Conjectures; and although this may be sufficient for any Mathematician, yet for your greater Satisfaction, I have thought proper to give two or three more evincing Arguments, to confirm these important Discoveries.   *Democritus*, as I have said before, believed them to be Stars long before Astronomy reaped any Benefit from the improved Sciences of Optics ; and saw, as we may say, through the Eye of Reason, full as far into Infinity as the most able Astronomers in more advantageous Times have done since, even assisted with their best Glasses: And his Conjectures are almost as old as the philolaic System of the Planets itself ; the Construction of which, though attempted by many, none have ever yet been able to confute.

The Light which naturally flows from this Crowd of radiant Bodies is mixt and confused, chiefly occasioned by the Agitation of our Atmosphere, and from a Union of their Rays of Light, by a too near Proximity of their Beams, altogether they appear like a River of Milk, but more of a pelucid Nature, running all round the starry Regions.

For

* *Plutarch (in Placitis Philosoph.)*

PLATE XIV.

For in the azure Skies its candid Way
Shines like the dawning Morn, or clofing Day.

There are alfo many more fuch luminous Spaces to be found in the Heavens of the fame Nature with thefe, which we know to be Stars; in particular the *Nebulæ*, or cloudy Star in the *Præfepe* of 36; a cloudy Star in *Orion* of 21; * a cloudy † Knot not far from this in the fame Afterifm of 80; in one Degree of the fame Conftellation 500, and in the whole Form above ‡ 2000. All of which are great Confirmations of the Truth of our Affertion, *i. e.* that this Zone of Light proceeds from an infinite Number of fmall Stars. Here it will not be amifs to obferve, that it has been conjectured, and is ftrongly fufpected, that a proper Number of Rays, meeting from different Directions, become Flame; and that hence it may prove not the Sun's real Body which we daily fee, but only his inflamed Atmofphere. I begin to be of Opinion, and I think not without Reafon, that the true Magnitude of the Sun is not near what the modern Aftronomers have made it; and that it may not poffibly be much above two Thirds of what it appears to us; I don't mean that this Expanfion of the folar Flame is any Part of that dilated Light mentioned by Sir *Ifaac Newton*, and conceived to be round all light Bodies in general; but you may confider it as not much differing from it, not of an unlike Nature, only greater in Degree, and peculiar to the Sun and Stars, who are all, as has been before in a manner demonftrated to be actually Globes of Fire.

This, tho I prefume to call it at prefent only meer Hypothefis, will in a great meafure account for the exceffive Changes in the Conftitution of our Air and Atmofphere, which we often find very unnatural to the Seafon; alfo be a Means perhaps of reconciling the vaft Difproportion fo very remarkable betwixt the Sun and the leffer Planets, and many other Circumftances in the Syftem of no fmall Confequence in Aftronomy: One of which Particulars you have frequently expreffed a great Miftruft and Difapprobation of, as fufpecting fome kind of a Fallacy in the Computation; and the other is Matter of general Complaint, being by many attributed to a Change in the Direction of the Earth's Axis‖; and by fome, efpecially the Vulgar, to too near an Approximation of the Earth to fome one of the celeftial Bodies. But all this will very naturally be accounted for by the Levity, or expanding Quality of the Sun's circumambient

---

* Vide *Galilæo.*
† Betwixt the Sword and Girdle of *Orion.*
‡ Vide *Reitha.*
‖ Which, through Ignorance of the true Cafe, is commonly called a Shock, a Brufh, or Shove.

Flame,

Flame, or Atmofphere; and hence, according to its various State, being more condenfed, or rare, we may have Heat or Cold in the greateft Extream, and alternately fo, in a perpetual Viciffitude.

The Truth of this Doctrine will evidently appear from the Obfervations of the Sun's Diameter through the Year 1660, by the indefatigable *Mouton:* And, I muft own, I am not a little furprized to find that no Conclufions have been drawn from them of this Kind. I am perfwaded, if you once compare thofe Numbers, you will be very far from thinking this an improbable Suggeftion. But this Digreffion has led me a little too far from the *Via Lactea,* and too near home again ; I muft now think of returning to the Stars, and my next Endeavours muft be to give you fome Idea of the Number of them. Through very good Telefcopes there have been difcovered in many Parts of this enlightened Space and even out of it, feveral thoufand Stars in the Compafs of one fquare Degree ; in particular near the Sword of *Perfeus,* and in the Conftellations of * *Taurus* and *Orion.*

PLATE XV.

Reprefents the *Pleides,* a well known Knot of Stars in the Sign *Taurus,* as they appeared to me thro' a one Foot reflecting Telefcope: And *Plate* XVI. is a View of the *Perfides,* another furprizing Knot of Stars in the Conftellation *Perfeus,* exactly as they appear through a Tube of two convex Glaffes. There are alfo other luminous Spaces in the ftarry Regions, not unlike the Milky Way, which I have had no Opportunity of obferving ; fuch as the *Nebeculæ,* near the South Pole, called by the Seamen *Magellanic* Clouds ; and which likewife viewed through Telefcopes, prefent us with little *Nebulæ,* and fmall Stars interperfed : One of thefe Kind is fituated between *Hydrus* and *Dorado* ; and another, fomething lefs than this, betwixt *Hydrus* and the *Toucan.*

Now admitting the Breadth of the *Via Lactea* to be at a Mean but nine Degrees, and fuppofing only twelve hundred Stars in every fquare Degree, there will be nearly in the whole orbicular Area 3,888,000 Stars, and all thefe in a very minute Portion of the great Expanfe of Heaven. What! a vaft Idea of endlefs Beings muft this produce and generate in our Minds ; and when we confider them all as flaming Suns, Progenitors, and *Primum Mobiles* of a ftill much greater Number of peopled Worlds, what lefs than an Infinity can circumfcribe them, lefs than an Eternity comprehend them,

---

* *Galilæo* in one cloudy Star of this Conftellation, difcovered no lefs than twenty-one, and in that of the *Præfepe* thirty-fix.

PLATE. XV.

PLATE XVI.

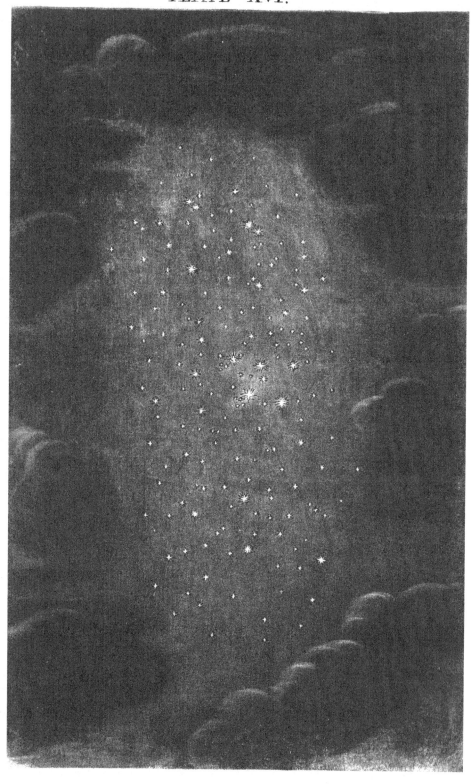

or lefs than Omnipotence produce and fupport them, and where can our Wonder ceafe?

In this Place perhaps I ought not to pafs over the aftonifhing Pheno-menon of feveral new Stars, &c. which have frequently appeared, and foon again vanifhed, in the fame Point of the Heavens. But as the Bu-finefs of this Theory is rather to folve the general, than any particula Phænomenon, I fhall only here by way of Note fubjoin a Table of fuch as has been regularly obferved, and by whom they were firft difcovered.

A Table of feveral new Stars, Nebulæ, and double Stars, &c.

Nomina Stellarum.	Obfervationum.
Septima Pleiadum	Loft after the burning of Troy, but now returned; fee Ricciolus.
A new Star appeared in Caffiopea, nearly in the fame Place with that of 1572.	Anno Dom. 945, bright as Jupiter; fee Ricciolus.
The new Star in Caffiopea's Chair.	Bright as Venus, from November 1572 to March 1574,
A new Star in Collo Ceti.	Of the 3d Magnitude, is faid to have appear'd periodically, feven Times in fix Years, i. e. every three hundred and thirteen Days: It was firft obferved in Auguft 1596, for two Months, by D. Fabricius
A new Star in the Swan's Neck,	Obferved by Kepler in 1600, of the third Magnitude, till the Year 1659; then gradually decreafing; in 1661 it difappeared; in 1666 it became vifible again, and is yet to be feen of the fixth Magnitude.
A new Star in the Right Foot of Serpen-tarius,	Bright as Venus from October 1604 to October 1605: fee Kepler.
A new Star in Andromeda's Girdle,	Seen by Simon Marius and Fabricius, Anno 1612.
A new Star in Antinous,	Seen by Juftus Byrgius.
A rew Star feen in the Whale,	In 1638, by John Procyclides Holuarda, of the third Mag-nitude, which difappeared periodically, every three hundred and thirty Days.
A new Star in the Fox's Head,	Of the third Magnitude, feen by Hevelius in July 1670, and till Auguft 1671, alfo from March 1672 to Septem-ber 1672
A new Star in the Swan's Neck.	This appear'd periodically every four hundred and four Days, and about fix Months at a Time; it was feen at its brighteft, September 10, 1714.

Of the Nebulæ, or Cloudy Stars.

Nebulofe in Orion's Sword.	
Nebulofe in Andromeda's Girdle	
Nebulofe in the Bow of Sagitarius,	Small, but very luminous.
Nebulofe in Centaurus,	Never feen in England.
A Nebulofe preceding the right Foot of Antinous,	Obfcure, but with a Star in the Middle of it.
Nebulæ in Dorfo Herculis,	Difcovered by Dr. Hally

Befides the Nebulæ, and new Stars, it appears from the ancient Catalogues of Hevelius, &c. that fome of the old ones have intirely vanifhed; in particular, one in the left Thigh of Aquarius, the contiguous one preceding in the Tail of Capricorn; the fecond on the Belly of the Whale; the firft of the unformed ones after the Scales of Libra, and feveral others. Many of the Stars alfo appear to be double, as the firft Star of Aries and Caftor; others triple, as one in the Pleiades; and the middle one in Orion's Sabre; and others again, quadruple, &c.

I would

I would now willingly help you to conceive the indefinite mutual Diſtance of the Stars, in order to give you ſome ſmall Notion of the Immenſity of Space ; but as this will be a Taſk merely conjectural, I ſhall only deſire you to believe it as far as your Reaſon will carry you, ſafely ſupported by an obvious Probability.

Perhaps it may be neceſſary here to acquaint you, that all the Stars are ſo far apparently of different Magnitudes, that no two of them are to be found in the whole Heavens exactly the ſame, either in Bigneſs or Brightneſs *. The largeſt we have ſufficient Reaſon to believe is the neareſt to us ; the next in Bigneſs and Brightneſs more remote ; and ſo on to the leaſt we ſee, which we judge to be the moſt remote of all.

The firſt Degree, or that of the largeſt Magnitude, we give to SYRIUS, the ſecond to ARCTURUS, the third to ALDEBARAN, the fourth to LYRA, the fifth to CAPELLA, the ſixth to REGULUS, the ſeventh to RIGEL, the eighth to FOMAHAUNT, and the ninth to ANTARUS: Theſe are all ſaid to be of the firſt Claſs ; and beſides which, there are at leaſt, within the Reach of our lateſt improved Opticks, nine more Denominations within the Radius of the viſible Creation.

Now, by the certain Return of the Comets, which we find are all governed by the Laws of this Syſtem, and ſuppoſed to be undiſturbed by any of the others, we cannot avoid concluding, if we conſider them at all to the Purpoſe, that the neareſt Stars cannot be leſs diſtant than twice the Radius of the greateſt Orbit belonging to the Sun. Moſt Mathematicians think this a great deal too near, as it muſt of courſe make all the Syſtems join, as in Contact ; and I think we may ſafely add, to ſeparate their Spheres of Attraction, at leaſt one Half of this Diſtance more, which will make in the Whole about four hundred and twenty Semi-orbits of the Earth, or 33,600,000,000 Miles. This even the ingenious Mr. *Huygins* endeavours to prove ſtill much too little, and his Arguments are ſuch as cannot eaſily be refuted. His Principle is grounded upon the known Laws of Analogy, as conſidered in the Proportion of light Surfaces, and is as follows. Having reduced the Sun's Diſk to the Appearance of the Star SYRIUS, by the Help of a ſmall Hole at the End of his Teleſcope, and comparing this Part of his Surface to the whole Diſk of the Sun, he infers that the Stars Diſtance to that of the Sun muſt be as 27,664 to 1. Hence *Syrius* from us will be nearly (avoiding Units) 2,213,120,000,000 Miles : But this I take to be as much too large as the former is too little ; yet, as

---

* A very little Knowledge in Opticks will render this indiſputable, and has been in a great meaſure demonſtrated before; 1. in the Great Dog ; 2. in Bootes ; 3. in the Bull; 4. in the Harp of *Apollo* ; 5. in *Auriga* ; 6. in the Lion ; 7. in *Orion* ; 8. in the Southern Fiſh ; 9. at the End of *Erridanus*.

Mr.

Mr. *Bradley* has, with some Shew of Reason, banished all the Stars out of the Sphere of Parallax, the last is the only Method we can possibly make use of with any kind of Confidence; and Sir *Isaac Newton* endeavours to recommend it with great Force of Argument, as the only probable Means by which we can give any tolerable Guess at these immense Measurements of Space.

To moderate the Matter then if you please, allow me but to make use of a Mean betwixt the two fore-mentioned Numbers; and we may take it for granted, a Distance sufficiently exact, to suit all our Wants in the present Case, namely, to give a very tolerable Idea of the Extent of the visible Creation, which is all I propose in this Place to attempt; but I mean to be much more exact in another.

Now as the Distance from the Sun to the Earth is so small in Proportion to the Distance of the Stars from us, and from one another, we may very well consider the Sun as the Center of our Station, or Position in the general System or Frame of Nature. And as the Stars are very visible thro' good Telescopes, to the ninth or tenth Magnitude, if we multiply the primary Distance of *Syrius*, or of any other of his Class, by this Number of common intermediate Spaces, the Product will be equal to the Radius of the visible Creation to the solar Eye; which, by this Rule, you will find in capital Numbers to be * nearly 6,000,000,000,000 Miles, taking in a Star of the sixth Magnitude, and to a Star of the ninth, 9,000,000,000,000 Miles: But this Computation supposes a mean common Distance of the Stars in a sort of Syzygia, or Direction of a Right Line, which is not the real Case; for the Stars cannot be supposed to diminish in a proportional Magnitude by any mathematical *Ratio*, but by some geometrical, or rather musical one; for Instance, if the Distance of a first be 3, that of a second should be about 5, and of a proportional Third 8,333, &c. *ad infinitum:* But as their true proportional Distance is unknown, the above will be sufficient for our present Purpose; which is only to shew, without Exaggeration, the Space we now are truly sensible of.

This I have here considered more extensively, to obviate all Objections that you may make to the Probability of the general Motion of the Stars, by shewing no Difficulty can possibly arise from their apparent Proximity, Number, or irregular Distribution: Their Distances being so immensely large, no Disorder or Confusion can be supposed in any Direction of them, or Motion whatever. The greatest Distance of the Planets, which all move undisturbed round the Sun, is about three hundred and fifty-three Million of Miles: But the least Distance of one Star from another, is

---

* If the Distance of the Sun and Earth is found too much, which I must own I have a violent Suspicion of, these Numbers must be reduced in like Proportion.

upwards

upwards of two thousand eight hundred and thirty-two Times that Distance, or one Million of Millions of Miles: And as no sensible Disorder can be observed amongst the solar Planets, what Reason have we to suppose any can be occasioned amongst the Stars, or that a general Motion of these primary Luminaries round a common Center, should be any way irrational, or unnatural?

What an amazing Scene does this display to us! what inconceivable Vastness and Magnificence of Power does such a Frame unfold! Suns crowding upon Suns, to our weak Sense, indefinitely distant from each other; and Miriads of Miriads of Mansions, like our own, peopling Infinity, all subject to the same Creator's Will; a Universe of Worlds, all deck'd with Mountains, Lakes, and Seas, Herbs, Animals, and Rivers, Rocks, Caves, and Trees; and all the Produce of indulgent Wisdom, to chear Infinity with endless Beings, to whom his Omnipotence may give a variegated eternal Life.

The astonishing Distance of the starry Mansions undoubtedly was design'd to answer some wise End: One Consequence is this, and probably is not without its Use: To every Planet of the same System, the same sidereal Face of Heaven appears without the least Degree of Change; and as the remotest Regions upon Earth see the same Moon and Planets, so also the Inhabitants of the most distant Planets in ours, or in any other System, see the same Forms and Order of the Stars in common with the rest. The whole Sphere of Heaven being common and unchangeable through all their various Revolutions.

Thus those (the People) in the Planet *Venus* will see the Constellation of *Orion* just as we do, and the People in the Planet *Saturn*, much farther still removed, alike will view this Constellation in all respects the same; here then, (in the System of the Sun) the Eye removed from us must only hope to find a new Earth surrounded with the same sort of Sky: But Beings in another System, behold not only a new Heaven above, but also new Earths below; and all the Frame of Nature to them puts on a new Dress, new Signs, new Seasons, and new Planets roll, and a new Sun renews the Day.

The Heathen Fables here are all erased with all the Immortality of their vain earthly Gods and Heroes; *Perseus* and *Alcides* are no more, and both the *Bears* are vanished; the *Pleiads* and the *Hyads* join, and shining *Leo*, though boasting two Stars of the first Magnitude with us, there no where can be found, lost in the common undistinguished Herd. But still Astronomy will exist, and new-framed Forms may fill the varied Scene.

Perhaps you may expect that I should here give you my Conjectures of what sort of Beings may be supposed to reside in the *Ens Primum*, or *Sedes*
Beatorum

Beatorum of the known Univerfe, whether mortal, immortal, or Creatures partaking in fome Degree of the Properties of both; as fuch may be conceiv'd to change their Natures and States, without a total Diffolution of their Senfes by Death : And farther, it may poffibly be judged unpardonable in me not to point out every bleffed Abode, fuited to the Virtues, and all the various States an immortal Soul may be tranflated to ; but this is a Tafk above the human Capacity, or is the pure Province of Religion alone ; the Bufinefs of a Revelation rather than Reafon to difcover. Befides, it is enough for the prefent Purpofe, to prove, that Miriads of celeftial Man-fions, are to be difcovered within our finite View, and by a kind of ocular Revelation, which vifibly extends the human Profpect, as it were, far be-yond the Grave. It matters not whether a Race of Heroes fill thefe Worlds, or a Tribe of happy Lovers people thofe ; whether a Peafant in the Realms of *Orion* fhall ever become a Prince in the Regions of *Arctu-rus*, or a Patriarch in *Procion*, a Prophet in the *Precepæ*. Not to mention all the Stages human Nature may, or have been deftined to in any one World, as believ'd by the ancient Philofophers, befides the final Coalition of all Beings much more naturally to be expected in the *Sedes Beatorum*.

I fay, whatever our Cafe may be with regard to thefe *Queries* and Futurity, the Plan and Principles of this Theory will not be at all changed by it, fince what it is chiefly founded upon may be clearly demonftrated, fo clearly and inconteftably, that, with the Reverend Dr. *Young*, we may juftly conclude,

Devotion! Daughter of Aftronomy!

and affirm with him alfo, That,

An indevout Aftronomer is mad.

But I find what I at firft propofed will prove too long for this Letter However, I will endeavour to reward your Patience in my next, and continue, &c.

LETTER

LETTER THE SIXTH.

Of General Motion amongst the Stars, the Plurality of Systems, and Innumerability of Worlds.

S I R,

SINCE my laſt, you'll find by this, ſpeaking in the Stile of *Kercher*, that I have been very far from home, round almoſt the viſible Creation. I have indeed applied myſelf very cloſely to tranſcribe my Thoughts to you upon the old Subject the *Milky Way*, which my former Letter left imperfected. To return then to the Theory of the Stars, and that yet unreconciled Phænomenon ; let us reaſon a little upon the viſible Order of the Stars in general, and ſee what Concluſions can be drawn from what every Aſtronomer knows of them, and cannot be diſputed.

First then, that the Stars are not infinitely diſperſed and diſtributed in a promiſcuous Manner throughout all the mundane Space, without Order or Deſign, is evident beyond a Doubt from this vaſt collective Body of Light, ſince no ſuch Phænomenon could poſſibly be produced by Chance, or exhibited without a deſigned Diſpoſit on of its conſtituent Bodies.

If any regular Order of the Stars then can be demonſtrated that will naturally prove this Phænomenon to be no other than a certain Effect ariſing from the Obſerver's Situation, I think you muſt of courſe grant ſuch a Solution at leaſt rational, if not the Truth ; and this is what I propoſe by my new Theory.

To a Spectator placed in an indefinite Space, all very remote Objects appear to be equally diſtant from the Eye ; and if we judge of the *Via Lactea* from Phænomena only, we muſt of courſe conclude it a vaſt Ring of Stars, ſcattered promiſcuouſly round the celeſtial Regions in the Direction of a perfect Circle.

But when we conſider the explanick Poſition of many other Stars, all of the ſame Nature, and not leſs numerous, together forming the great Sphere of Heaven, we generally find ourſelves quite at a Loſs how to reconcile the two apparent Claſſes ; and I know none who have ever been ſucceſsful enough to reduce them to any one general Order.

You'll

You'll fay probably how fhall we make this chaofic Difpofition of the primary Luminaries agree with the fecondary Laws, and the juft Harmony obferved in the third * Creation, &c.

The Work now you fee is undertaken, and chiefly at your own Requeft, therefore I have a Right to expect you'll be very indulgent to the Author, and pafs over all his Faults, and allow him free Argument in Purfuit of thefe important Truths, which will in the End open perhaps a much wider Field of Contemplation to us, than at firft could be fuppofed to be intended by the *Genefis* of *Mofes*.

That Defcription of the Beginning of Nature is not without its Beauty and Noblenefs, fuitable to the Dignity both of the Author and Subject. But fhould we even in this knowing Age of the World pretend to account for the Original of Things, as *Mofes* to fupport his believed divine Legation, was obliged in fome meafure to do, we fhould foon be reduced to talk in the fame Stile, and perhaps with lefs Probability, than then at leaft appeared in his elegant Account of the Origin of the Univerfe, efpecially if we do but confider, that what he wrote, was only to the Senfes of a People who had not yet learnt to make ufe of their Reafon any other way, but from the Appearance of Things, and upon a Subject too fublime for vulgar Capacities in any Age, and had only been attempted in the deepeft Learning of *Egypt*, which, he though well acquainted with, the Generality of them were totally Strangers to.

In the firft Place it muft be granted, that the Stars being all of the fame Nature, are either all immoveable, or all fixed, that is all governed by one and the fame Principle.

Now to fuppofe them all fixed, and difperfed in an endlefs Diforder thro' the infinite Expanfe, which has long been the Opinion of many very able Aftronomers amongft the Antients, and even now received by too many of the Moderns, implies an Inactivity in thofe vaft and principal Bodies, fo much the Reverfe of what may be expected, and what we daily obferve through all the reft of their Attendants, namely, their own refpective Satellites, that we cannot poffibly upon any rational Grounds, advance one fingle Argument to fupport fo much as a Conjecture towards it, without betraying the greateft Simplicity, and next to an Affirmation reduce the whole Frame of Nature, and all corporeal Beings to a wild unmeaning Chance, arifing from an unnatural Difcord and Confufion.

For upon the Principles of Locality and Materiality, you having allowed me the Ufe of my Senfes and Reafon, as abfolutely neceffary towards conceiving any Idea of our prefent State, or of Futurity: Upon

---

* The Moon, Satellites of *Saturn* and *Jupiter*, &c.

thefe

these Principles I say, unless our Faculties are useless, if there are no other Bodies or Beings in the Universe than what we see, and are now sensible of, we must now at the Height of this our present State, be as near Perfection as we can reasonably expect, and as such ourselves the supreme Beings of all Beings. To what End then do we form Ideas of a succeeding Life, where a more exalted State cannot be hoped for.

How absurd and impious this is I leave to your own Reason and Reflection: This is the fatal Rock upon which all weak Heads and narrow Minds are lost and split upon, consequently ought to be the most carefully avoided, not only as the Nurse of Atheism, but as the dreadful Father of Despair: " For, say they, these unhappy Wretches, to be always the " same, is inconsistent with a Change ; and to be less than what we are, " any where hereafter, is full as difficult to conceive as to be more." Thus unless we admit of superior Seats and much more glorious Habitations than these we are sensible of, we strike at the very Root of a fair flourishing Tree of Immortality, and must become Authors of our own Despair. I have often wonder'd how thinking Men could possibly fall into so gross an Error, as that of a Spirit's Annihilation ; and I should be glad to ask one of those fruitless Students, whether, upon the Evidence of our present Being, it is not much more rational, to hope for a future, than to expect a *Ne plus ultra* upon no Evidence at all. The Affirmative is certainly much more natural to be conceiv'd than the Negative. But if Chance were the Case, and that Chance produced all these regular and wondrous Works, tis to be wished at least, that Chance might do the same again ; and if not Chance, of course an eternal Direction: But Chance only can effect Disorder, Discord, and Confusion ; *ergo*, the visible Harmony and Beauty of the Creation declare for a Direction ; and this must of Consequence, from its perfect Nature, proceed from the Wisdom and Power of an eternal Being, *God of Infinity*, the Author of all Ideas : And if this primitive Power produced us his Creatures from nothing, nothing can be wanting to revive our Frames again ; and if from something, that something must remain to establish us in a future Life. But to return, how absurd it is to suppose one Part of the Creation regular, and the other irregular, or a visible circulating Order of Things, to be mixed with Disorder, and circumscribing Part of an endless Confusion, is obvious to the weakest Understanding, and consequently we may reasonably expect, that the *Via Lactea*, which is a manifest Circle amongst the Stars, conspicuous to every Eye, will prove at last the Whole to be together a vast and glorious regular Production of Beings, out of the wondrous Will or Fecundity of the eternal and infinite *one* self-sufficient Cause ; and that all its Irregularities are only such as naturally arise from our excentric View : To demonstrate which

PLATE XVII.

which abſolutely and inconteſtibly, we ſhall only want this one *Poſtulata* to be granted, *viz. That all the Stars are, or may be in Motion:* This, if one may be allowed to judge of the Whole by the Similitude and Government of its Parts, I am perſwaded you will think a very reaſonable Aſſumption; but that you may imbibe a good Opinion of this Aſſumption, and entirely come into this much better to be wiſhed Hypotheſis, I would have you conſult theſe following Arguments.

Firſt, it is allowed, as I have endeavoured to ſhew, by all modern Philoſophers, that the Sun and Stars are all of the ſame or like Nature; conſequently, that the Stars are all Suns, and that the Sun himſelf is a Star.

PLATE XVII.

Repreſents a kind of perſpective View of the viſible Creation, wherein A repreſents the Syſtem of our Sun, B, that ſuppoſed round *Syrius*, and C, the Region about *Rigel*. The reſt is a promiſcuous Diſpoſition of all the Variety of other Syſtems within our finite Viſion, as they are ſuppoſed to be poſited behind one another, in the infinite Space, and round every viſible Star. That round every Star then we may juſtly conjecture a ſimilar Syſtem of Bodies, governed by the ſame Laws and Principles with this our ſolar one, though to us at the Earth for very good Reaſons inviſible *. Secondly,

The Sun is alſo obſerved to have a Motion round his own Axis in about twenty-five Days. Now, ſince all the other † Planets which move in Orbits round him, and are within our Obſervation, are found to have a like Rotation round their Axis, may we not as reaſonably imagine, that that Power which was able to give the Sun a Motion round his Axis, could and would at the ſame time, with adequate Eaſe, give him alſo an orbitular one? and why not, ſince no progreſſive Mutability can either take from, or diſturb the boundleſs Property of an Infinity; and beſides, ſeeing to imagine him at reſt, is to impoſe ſuch an unnatural Stagnation upon the eternal Faculty, quite repugnant to that imparable Power which we ſuppoſe ſtands in need of neither Sleep nor Reſt?

'Tis true, the Sun may be ſaid to be the Governor of all thoſe Bodies round him; but how? no otherwiſe than he himſelf may be governed by a ſuperior Agent, or a ſtill more active Force; and methinks it is not a

---

* *Anaximines* believed the Stars to be of a fiery Nature; and that there were certain terreſtrial Bodies that are not ſeen by us, carried together round them. *Stob. Ecl. Phyſ.* cap. 25. *Pythagoras* affirmed, that every Star is a World, containing Earth, Air, and Æther.

† *Saturn, Jupiter, Mars, Venus,* the Earth, Moon, and *Mercury.*

little

little abſurd to ſuppoſe he is not, ſince we have diſcovered by undoubted Obſervations, that the ſame gravitating Power is common to all ; and that the Stars themſelves are ſubject to no other Direction than that which moves the whole Machine of Nature.

Thirdly, From many Obſervations of the polar Points, and the Obliquity of the Earth's Equator to the Plane of her ſolar Orbit compared together, the Sun is very juſtly ſuſpected to have changed his ſidereal Situation ; and this muſt either ariſe from a Change in the Poſition of the Earth's diurnal Axis, or from a Removal of the Sun himſelf, out of the primitive Plane of the *Orbis Magnus*. I believe you are ſo much of a Mathematician, as to know that if either of theſe Facts be allowed, the Conſequence I want will follow. I ſhall not therefore here enter into any farther Diſpute about it ; but I think it will be neceſſary to ſubmit ſome Obſervations to your Conſideration, that may convince you that there is a Motion ſomewhere to be thus diſcovered, and whether in the Sun, or in the Stars, or in both, I leave to your own Determination, but to aſſiſt your Imagination, I refer you to

PLATE XVIII.

The Globe S is here ſuppoſed to repreſent the Sun, having changed its Situation by a local Motion from A to C, and B repreſents the Globe of the Earth in a permanent Poſition, with its principal Points and Circles, reſpecting the primitive Plane A, B, K. Now in Conſequence of the Angle of Variation, A, B, C, it evidently appears that a new ecliptic Plane, will be produced, as C, B, and alſo a Variation in the greateſt Declination of the Sun, North and South from the Line of the *Equator* D, L. Hence, as in this Figure, the Obliquity of the Poles P, N, and G, F, will naturally decreaſe, and is ſhewn in Quantity by the Line of Aberration H, I.

Here follows a Table of the Change obſerved in the Obliquity of the Ecliptic by Aſtronomers of different Ages.

A Table of the Obliquity of the Ecliptic.

Ante Chriſti		°	′
124	ARATO - - - - - - - - - - -	24	00
——	HIPARCHUS - - - - - - - - - -	23	51 $\frac{1}{3}$
127	ERATOSTHENES - - - - - - - - - -	23	51 $\frac{1}{3}$

Anno

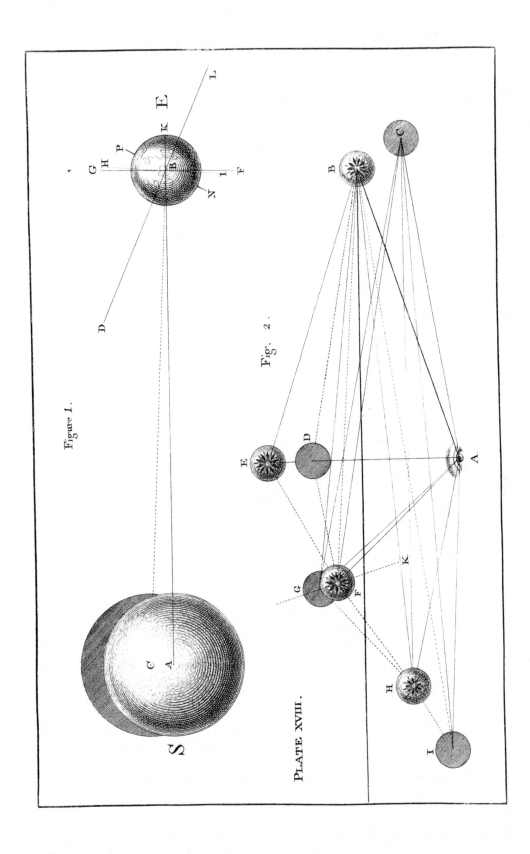

Figure 1.

S

Fig. 2.

PLATE XVIII.

Anno Dom.			o	'
140	PTOLOMY	- - - - - - - - - -	23	51 $\frac{1}{3}$
749	ABATEGNIUS	- - - - - - - -	23	35 $\frac{1}{2}$
1070	AIRAHEL	- - - - - - - -	23	34
1140	ALOMEAN	- - - - - - - -	23	33
1300	PROFATIOGRAD	- - - - - - -	23	32
1458	PURBACCHIO	- - - - - - - -	23	29 $\frac{1}{2}$
1490	REGIOMONTAUS	- - - - - - -	23	30
1500	COPERNICUS	- - - - - - - -	23	28 $\frac{1}{2}$
1592	TYCHO BRAHE	- - - - - -	23	21 $\frac{1}{2}$
1656	CASSINI	- - - - - -	23	29 $\frac{1}{2}$

Now sure, if we consider this continual Decrease of the Sun's Declination, which can proceed from no other Cause than that of his having moved out of the primitive Plane; we need make no great Difficulty thus far, to think our Conjectures not irrational.

The following is a Citation from Dr. *Edmund Hally*, Astronomer-Royal. See *Philosophical Transactions*, Nᵒ. 355. p. 736.

" But while I was upon this Enquiry (*of the Obliquity of the Ecliptic*) I was surprized to find the Latitudes of three of the principal Stars in the Heavens, directly to contradict the supposed greater Obliquity of the Ecliptic, which seems confirmed by the Latitudes of most of the rest; they being set down in the old Catalogues, as if the Plane of the Earth's Orbit had changed its Situation amongst the fixed Stars, about 20' since the Time of *Hipparchus*, particularly all the Stars in *Gemini* are put down, those to the Northward of the Ecliptic, with so much less Latitude than we find, and those to the Southward, with so much more southerly Latitude; and yet the three Stars *Palilicium*, *Sirius*, and *Arcturus*, do contradict this Rule: For by it, *Palilicium*, being in the Days of *Hipparchus*, in about 10 gr. of *Taurus*, ought to be about 15' more southerly than at present, and *Sirius* being then in about 15 gr. of *Gemini*, ought to be 20' more southerly than now; yet *Ptolomy* places the first 20', and the other 22' more northerly in Latitude than we now find them: Nor are these the Errors of Transcribers, but are proved to be right by the Declination of them set down by *Ptolomy*, as observed by *Timocharis*, *Hipparchus*, and himself; which shew, that these Latitudes are the same as those Authors intended. As to *Arcturus*, he is too near the Equinoctial Colure, to argue from him concerning the Change of the Obliquity of the Ecliptic; but *Ptolomy* gives him 33' more North Latitude than

than he is now found to have; and that greater Latitude is likewife confirmed by the Declinations delivered by the abovefaid Obfervations: So then thefe three Stars are found to be above half a Degree more foutherly at this Time than the Antients reckoned them. When, on the contrary, at the fame time, the bright Shoulder of *Orion*, has, in *Ptolomy* almoft a Degree more foutherly Latitude than at prefent, what fhall we fay then? It is fcarce to be believed, that the Antients could be deceived in fo plain a Matter, three Obfervers confirming each other. Again, thefe Stars being the moft confpicuous in Heaven, are in all Probability the neareft to the Earth; and if they have any particular Motion of their own, it is moft likely to be perceived in them, which in fo long a Time as eighteen hundred Years, may fhew itfelf by the Alteration of their Places, though it be intirely imperceptible in the Space of one fingle Century of Years: Yet, as to *Syrius*, it may be obferved, that *Tycho Brahe* makes him 2 Min. more northerly than we now find him; whereas he ought to be above as much more foutherly from his Ecliptic (whofe Obliquity he makes $2'\frac{1}{2}$ greater than we efteem it at at prefent) differing in the Whole $4'\frac{1}{2}$.

One Half of this Difference may perhaps be excufed, if Refraction were not allowed in this Cafe by *Tycho*; yet 2 Min. in fuch a Star as *Syrius*, is fomewhat too much for him to be miftaken in.

But a more evident Proof of this Change is drawn from the Obfervation of the Application of the Moon to *Pal.licium, An. Chrif.* 509. *Mar.* 11. when, in the Beginning of the Night, the Moon was feen to follow that Star very near, and feemed to have eclipfed it, ἐπέβαλλε γὰρ ὁ ἀστὴρ τῷ πᾶρα τὴν διχοτομίαν μέρει τῆς κύρτυς περιφειας τοῦ πεφωτισμένου μερους, *i. e. Stella appofita erat parti per quam bifecabatur limbus Lunæ illuminatus*, as *Bulliadus*, to whom we are beholden for this ancient Obfervation, has tranflated it. Now, from the undoubted Principles of Aftronomy, this could never be true at *Athens*, or near it, unlefs the Latitude of *Palilicium* were much lefs than we at this Time find it *.

The Motion of *Arcturus* feems further confirmed, from the Obfervations of *Tycho Hevelius* and *Flamftead*; for *Hevelius* fets down the Diftance of that Star from *Lyra* 4′ greater than *Tycho* had obferved it feventy-two Years before him, and *Flamftead* twenty-two Years after meafured

---

* Vide *Bulialdi Aftr. Philolaica*, p. 172.

† Thefe are the neareft and greateft of the fixed Stars, the Motion of the others not having been obferved, or being at too great a Diftance, are either imperceptible, or have not been taken notice of.

the Diſtance betwixt the ſame two Stars, ſtill 3′ greater than *Hevelius* found it ; ſo that if *Lyra* had ſtood ſtill all that while, there was an Appearance of *Arcturus*'s having gone 7′ out of his Place in the Space of an hundred Years. See Dr. *Long*'s Aſtronomy, p. 274.

It is further to be obſerved, in Confirmation of the Motion of one of theſe Stars, that *Flamſtead* found the Diſtance of *Arcturus*, from the Head of *Hercules* 3′ greater than it is ſet down by the Prince of *Heſſe* ; and that his Diſtance from the *Lion's Tail* was a little decreaſed with 5′ ½ leſs Latitude than *Tycho* had obſerved. Hence, to make theſe Obſervations agree, one or both of them muſt have moved together equal to 7′. This Change of Place, which is quite contrary to all known Cauſes proceeding from the Earth, muſt therefore be occaſioned either by the Motion of the Sun, or by a particular Motion of their own ; but if, amongſt themſelves, they muſt all move, and if all be in Motion, the Sun muſt alſo move.

If theſe Obſervations, delivered down to us by very able Aſtronomers, be either true or near it, as great Allowances have been made for the Ignorance of the Ages in which they were taken, and the Inaccuracy of the Inſtruments, we may naturally conclude, that theſe Stars muſt have a Motion ; and if they move, as has been before obſerved, the Sun muſt alſo ; hence he cannot now be in the original Plane of the Earth's annual Direction, or at leaſt in the ſame identical Place he was at firſt poſſeſſed of : And if ſo, the Stars muſt alſo have the like Motion, though in different Directions, and all may thus be governed by the ſame impulſive Power.

To illuſtrate this primitive Motion of the Stars, and at the ſame time to ſhow that the Variety which appears in the Quantity of Motion can be no Objection to it,

See PLATE XVIII. *Fig.* 2.

Where A repreſents the Eye of an Obſerver, and B, E, F, H, various Syſtems, moving in different Directions thro' the mundane Space; it is evident that the Sphere B, having moved from C, and that of E, not having appeared to move at all, there muſt be a ſenſible Change in the new Poſition of theſe two Syſtems to one another, and ſo of the reſt; and tho' the apparent Motion of H, be much more than that of F, from the Point A, yet from C, they will appear leſs different, and from B, they will appear nearly equal. And farther, as the Direction from H, is in the Line I, H, and that of F, in the Line K, G, thoſe two Syſtems will appear to approximate, and the Magnitude of the Star in the firſt will be increaſed, creafed, and in the latter diminifhed. Thus, many of the Stars in the oldeft Catalogues, which were faid to be of the fecond Magnitude, are now become of the firft, and feveral of the firft are now judged to be of the fecond, &c.

But as this apparent Motion of the Stars at the Earth, muft, from its Nature, be very fmall, fo as fcarce to be difcovered in fome of them in lefs than an Age, with any Inftrument by the niceft Obferver, I judge it will be extremely proper in this Place to propofe fome Method, by which, in procefs of Time, the Truth of the Theory may be afcertained. The Way I think moft likely to fucceed is this.

P L A T E XIX.

Is a Plan of the principal Stars that form the PLEIADES, correctly taken by a Combination of Triangles, as in the Figure, from whence it will naturally follow, all the whole Form being comprehended in much lefs than one Degree. That the moft minute local Motion in any one of thofe Stars in a very few Years, will be made fenfible to an Eye at the Earth. For Inftance, if any of the Stars that form the Letter A, or T, within the Term of ten or twenty Years, be found in the leaft to deviate from the Lines of their prefent Pofition and Direction, it will be evident beyond a Contradiction, that they have a Motion amongft themfelves, and fince at fuch a Diftance they cannot poffibly be affected by the Earth, it muft be a Motion of their own ; and thus if any one can be proved, to change its Situation, with regard to the reft, we can have no new Difficulty in concluding that they all may do the fame.

Thus if any of the regular Triangles M B Z, Z P H, A Z M, T A Г, or п O I, &c. in due Time be carefully noted, we may venture to fay with great Safety, that the thoufandth Part of a Degree will be plainly difcovered.

P L A T E XX.

Is a true Plan and Combination of the principal Stars that form the PERSEDES, in which other Obfervations may be made in a different Part of the Heavens, and perhaps with an Opportunity of being ftill more exact, the Areas of thefe Triangles, particularly that of o I K, and thofe of ρ and δ, being much lefs than the former, where the leaft Alteration poffible muft render them fenfibly diftorted. But here it muft be confidered, that the real Motion of the Stars, as well as their apparent, may be, and in

all

PLATE. XIX.

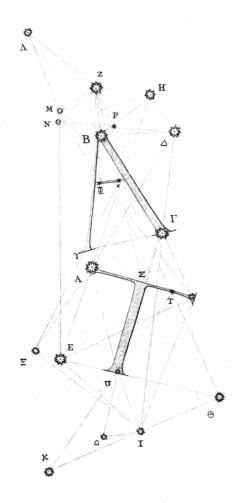

PLATE. XX.

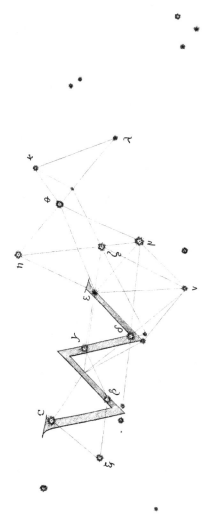

all Likelihood, is extreamly flow, for the moſt minute, viſible, local Motion, will anſwer all the Purpoſes we know in Nature, and the greateſt ſeems to be that of the projectile, or centrifugal Force, which not only preſerves them in their Orbits, but prevents them from ruſhing all together, by the common univerſal Law of Gravity, which otherwiſe, as a finite Diſtribution of either regular or irregular Bodies, they muſt at length do by Neceſſity.

I muſt now inform you, that the above Obſervations were compleated in the AUTUMN SEASON, 1747, and were taken by myſelf; the Letters A, T, in *Plate* XIX, and the W in the XXth, as you may ſee, having a very near Reſemblance, or Similitude, to the Order theſe Stars are found to be in, together with the *Greek* Alphabet, I judged neceſſary, by way of *Aſteriſm* and *Nomenclatura*, in caſe ſuch ſhould be wanted, as *Data* in future Diſcoveries.

I come now to the principal Point in Queſtion, which is to find a regular Diſpoſition of the Stars amongſt themſelves, which will naturally ſolve both their general and particular Phænomena, eſpecially the *Nebula* and *Milky Way*.

I am now, &c.

I   LETTER

LETTER THE SEVENTH.

The Hypothesis, or Theory, fully explained and demonstrated, proving the sidereal Creation to be finite.

SIR,

I KNOW you are an Enemy to all Sorts of Schemes where they are not absolutely necessary, and may possibly be avoided; and for that Reason I have purposely omitted many geometrical Figures, and other Representations in this Work, which might have been inserted and in some Places, especially here I might have introduced Diagrams, perhaps more explicit than Words; but as you have frequently observed, they are only of Use to the few Learned, and contribute more to the taking away the little Ideas and Knowledge the more ignorant Many may be endued with, by a prejudicial Impression of imperfect Images, rather than the adding any new Light to their Understanding, I have purposely avoided, as much as possible, both here and every where, all such complex Diagrams as might be in Danger of betraying any the least such conscious Diffidence in you, arising from the Want of a proper *Precognita* in the Sciences.

This Imperfection, much to be lamented, as greatly to the Disadvantage of all mathematical Reasoning, I would willingly always prevent, in my Readers, and to chuse in my Friend; I shall therefore content myself with referring you to a few orbicular Figures, concave and convex, as may best suggest to your Fancy the simplest Way, a just Idea of the Hypothesis I have fram'd, and naturally enough I hope, render my Theory so intelligible, as to help you sufficiently to conceive the Solution aimed at, of the important Problem I have attempted.

As I have said before, we cannot long observe the beauteous Parts of the visible Creation, not only those of this World on which we live, but also the Myriads of bright Bodies round us, with any Attention, without being convinced, that a Power supreme, and of a Nature unknown to us, presides in, and governs it.

The

> The Courfe and Frame of this vaſt Bulk, diſplay
> A Reaſon and fix'd Law, which all obey.
>
> SHER. MANILIUS.

And notwithſtanding the many wonderful Productions of Nature in this our known Habitation, yet the Earth, when compared with other Bodies of our own Syſtem, ſeems far from being the moſt conſiderable in it; and it appears not only very poſſible, but highly probable, from what has been ſaid, and from what we can farther demonſtrate, that there is as great a Multiplicity of Worlds, variouſly diſperſed in different Parts of the Univerſe, as there are variegated Objects in this we live upon. Now, as we have no Reaſon to ſuppoſe, that the Nature of our Sun is different from that of the reſt of the Stars; and ſince we can no way prove him ſuperior even to the leaſt of thoſe ſurpriſing Bodies, how can we, with any Shew of Reaſon, imagine him to be the general Center of the whole, *i. e.* of the viſible Creation, and ſeated in the Center of the mundane Space? This, in my humble Opinion, is too weak even for Conjecture, their apparent Diſtribution, and * irregular Order argue ſo much againſt it.

The Earth indeed has long poſſeſſed the chief Seat of our Syſtem, and peaceably reigned there, as in the Center of the Univerſe for many Ages paſt; but it was human Ignorance, and not divine Wiſdom, that placed it there; ſome few indeed from the Beginning have diſputed its Right to it, as judging it no way worthy of ſuch high Eminence. Time at length has diſcovered the Truth to every body, and now it is juſtly diſplaced by the united Conſent of all its Inhabitants, and inſtead of being thought the moſt majeſtick of all Nature's lower Works, now rather diſgraces the Creation, ſo much it is reduced in its preſent State from what it had Reaſon to expect in the former.

Now it is no longer the only terreſtrial Globe in the Univerſe, but is proved to be one of the leaſt Planets of the ſolar Syſtem, and ſurprizingly inferior to ſome of its Fellow Worlds. The Sun, or rather the Syſtem, has almoſt as long uſurped the Center of Infinity, with as little Pretence to ſuch Pre-heminence; but now, Thanks to the Sciences, the Scene begins to open to us on all Sides, and Truths ſcarce to have been dreamt of, before Perſons of Obſervation had proved them poſſible, invades our Senſes

---

* See the Zodaical Conſtellations, you'll find that in ſome Signs there are ſeveral Stars of the firſt, ſecond, and third Magnitude, and in many others none of theſe at all.

I 2

with

with a Subject too deep for the human Understanding, and where our very Reason is lost in infinite Wonders. How ought this to humble every Mind susceptible of Reason!

In this Place, I believe, you will pardon a Digression; which, in Answer to Part of your last Letter, I judge will not be very impertinent, tho' perhaps just here I cannot so well justify it.

Your late Conversation with our Friend Mr. * * *, I am persuaded, must have been very entertaining; but I cannot help thinking his Reflections upon the Wonders of Nature and the Wisdom of Providence, though I must allow them all to be very just and curious, instead of elevating the Mind to the Pitch he would have it, rather as considered above, depress it below the proper, nay I might say necessary, Standard of human Ideas.

This, probably, you'll say is an odd Turn, and may want some Explanation, since every Object in the Chain of Nature, must of Force be granted, a Subject worthy of our Speculations, being all together made, as in the Maximum of Wisdom: But what I mean is this, since nothing is more natural for Beings in every State in search after their own Advantages, and the Enlargement of their Ideas to look upward, sure it may be presumed, that Time may be misspent, if not lost in inspecting too narrowly Things so little benefical in States below us; as Mr. *Pope* says,

> Why has not Man a microscopic Eye?
> For this plain Reason, Man is not a Fly.
> Say what the Use, were finer Opticks given,
> To inspect a Mite, not comprehend the Heav'n.
>
> *Essay on Man.*

Amusement alone can never be supposed to be the sole End of human Life, where even true Happiness is a Thing we rather taste than enjoy. The Mind we find capable of much more rational Pleasure than can possibly fall within the Reach of human Power, either to promise or procure it; but then this very Defect in our present State of Existence affords us no less than a moral Assurance, that some where in a future, we may, if we please, be entitled to the very *Plenum* of all Enjoyments.

The peculiar Business then of the human Mind naturally precedes its Amusements, as evidently ordained to soar above all the inferior Beings of this World; and however our Natures may, thro' Indolence, or thro' Ignorance, degenerate, that of the Man can never be supposed to sink into the Mole.

The properest Way then sure for Men to preserve their Pre-heminence over the Brute Creation, is to make use of that Reason and Reflection,

which

which fo manifeftly diftinguifhes their natural Superiority. A right Application of which, muft of courfe then direct us to a forward, rather than
a backward Search in the vaft vifible Chain of our Exiftence, which clearly
connects all Beings and States as under the Direction of one fupreme Agent.

This is all I would have underftood by the foregoing Pofition, which,
in one Word, implies no more than that the fublime Philofophy ought
in all Reafon to be preferred to the Minute; but I hope you will not infer
from this my feeming Partiality for the celeftial Sciences, that I mean to
infinuate, that the Study of terreftrial Phyficks is not a rational Amufe-
ment.

Mr. ***, you fay, feems to lament the Tafte of Mankind in general much
in the fame Degree as you do his I readily grant you ; a Man who can talk
fo well upon an Ant, might make a more entertaining Difcourfe upon the
Eagle; but I beg his Pardon, and though we are all too ready, and moft
apt to condemn all fuch Pleafures as vain or trifling, which we have no
Share in, or Tafte for ourfelves; yet I don't think it follows, that thofe ingenious Labours of his are ufelefs. The Pleafures arifing from natural
Philofophy are all undoubtedly great ones, whether we confider Nature in
her higheft, or in her loweft Capacity ; the Beauties of the Creation are
every Day varied to us below, as much they are every Night above, and in
both Cafes, through every Object, the Creator fhines fo manifeft, that we
may juftly confider him every where fmiling full in the Face of all his
Creatures, commanding as it were an awful Reverence, and Refpect, due
not only to his Omnipotency, but alfo to his infinite Goodnefs and endlefs
Indulgencies. This is the only Return our Gratitude can make for all thofe
Bleffings he daily beftows upon us, and to this great Author of her Laws,
Nature herfelf cries aloud through Myriads of various Objects, and after
her own expreffive and peculiar Manner, feems to command us with an
attractive Grace, to obferve her Sovereign, and admire his Wifdom. The
Majefty, Power, and Dominion of God is beft difplayed in the external
Direction of Things, his Wifdom and vifible Agency in the internal :
Hence, by proper Objects, felected from both, attended with juft Reflections, we may certainly raife our Ideas almoft to the Pitch of Immor-
tals ; but how far the human Imagination may poffibly go, or how much
Minds like ours may be improved, is a Queftion not eafily determined ;
but as natural Knowledge evidently increafes daily, and aftronomical Enquiries are the moft capable of opening our Minds, and enlarging our
Conception, of confequence they muft be moft worthy our Attention of
all other Studies. But of this I have faid enough, and think it is now
more than Time to attempt the remaining Part of my Theory.

When

When we reflect upon the various Aspects, and perpetual Changes of the Planets, both with regard to their * heliocentric and geocentric Motion, we may readily imagine, that nothing but a like eccentric Position of the Stars could any way produce such an apparently promiscuous Difference in such otherwise regular Bodies. And that in like manner, as the Planets would, if viewed from the Sun, there may be one Place in the Universe to which their Order and primary Motions must appear most regular and most beautiful. Such a Point, I may presume, is not unnatural to be supposed, altho' hitherto we have not been able to produce any absolute Proof of it. See *Plate* XXV.

This is the great Order of Nature, which I shall now endeavour to prove, and thereby solve the Phænomena of the *Via Lactea*; and in order thereto, I want nothing to be granted but what may easily be allowed, namely, that the *Milky Way* is formed of an infinite Number of small Stars.

Let us imagine a vast infinite Gulph, or Medium, every Way extended like a Plane, and inclosed between two Surfaces, nearly even on both Sides, but of such a Depth or Thickness as to occupy a Space equal to the double Radius, or Diameter of the visible Creation, that is to take in one of the smallest Stars each Way, from the middle Station, perpendicular to the Plane's Direction, and, as near as possible, according to our Idea of their true Distance.

But to bring this Image a little lower, and as near as possible level to every Capacity, I mean such as cannot conceive this kind of continued Zodiack, let us suppose the whole Frame of Nature in the Form of an artificial Horizon of a Globe, I don't mean to affirm that it really is so in Fact, but only state the Question thus, to help your Imagination to conceive more aptly what I would explain *. *Plate* XXIII. will then represent a just Section of it. Now in this Space let us imagine all the Stars scattered promiscuously, but at such an adjusted Distance from one another, as to fill up the whole Medium with a kind of regular Irregularity of Objects. And next let us consider what the Consequence would be to an Eye situated near the Center Point, or any where about the middle Plane, as at the Point A. Is it not, think you, very evident, that the Stars would there appear promiscuously dispersed on each Side, and more and more inclining to Disorder, as the Observer would advance his Station towards either Surface, and nearer to B or C, but in the Direction of the general Plane towards H or D, by the continual Approximation of the visual Rays, crowding together as at H, betwixt the Limits D and G, they must in-

---

* Not to mention their several Conjunctions and Apulces to fixed Stars, &c. see the State of the Heavens in 1662, *December* the first, when all the known Planets were in one Sign of the Zodiac, viz. *Sagittarius*.

fallibly

PLATE.XXI.

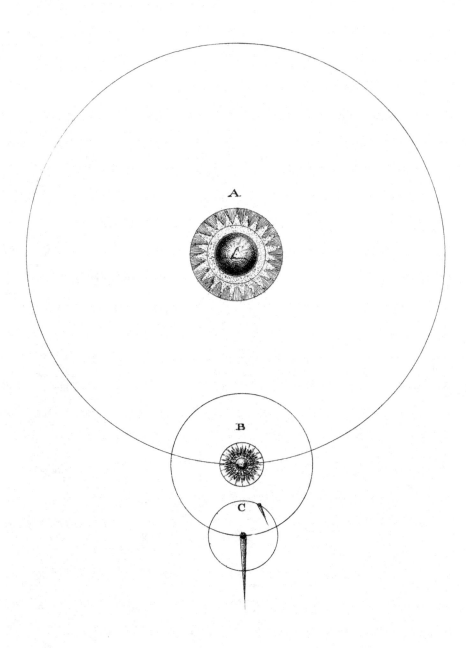

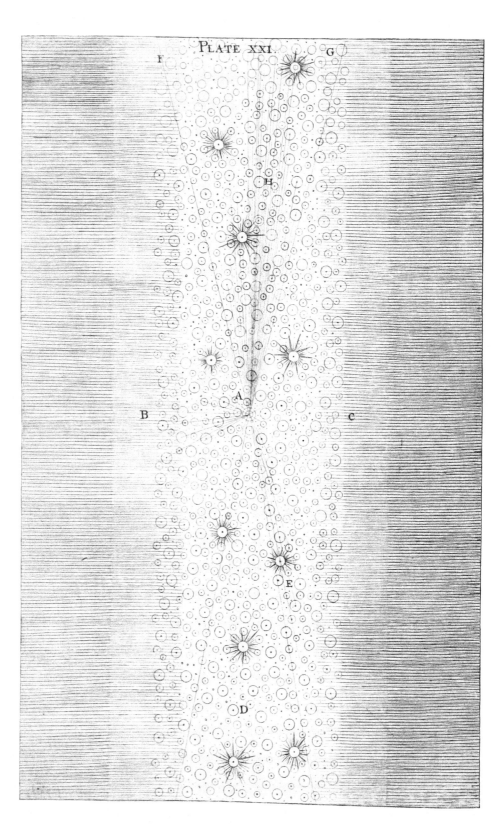

fallibly terminate in the utmoft Confufion. If your Opticks fails you be-fore you arrive at thefe external Regions, only imagine how infinitely greater the Number of Stars would be in thofe remote Parts, arifing thus from their continual crowding behind one another, as all other Objects do towards the Horizon Point of their Perfpective, which ends but with Infinity: Thus, all their Rays at laft fo near uniting, muft meeting in the Eye appear, as almoft, in Contact, and form a perfect Zone of Light; this I take to be the real Cafe, and the true Nature of our *Milky Way*, and all the Irregularity we obferve in it at the Earth, I judge to be intirely owing to our Sun's Pofition in this great Firmament, and may eafily be folved by his Excentricity, and the Diverfity of Motion that may naturally be conceived amongft the Stars themfelves, which may here and there, in different Parts of the Heavens, occafion a cloudy Knot of Stars, as perhaps at E.

But now to apply this Hypothefis to our prefent Purpofe, and reconcile it to our Ideas of a circular Creation, and the known Laws of orbicular Motion, fo as to make the Beauty and Harmony of the Whole confiftent with the vifible Order of its Parts, our Reafon muft now have recourfe to the Analogy of Things. It being once agreed, that the Stars are in Motion, which, as I have endeavoured in my laft Letter to fhew is not far from an undeniable Truth, we muft next confider in what Manner they move. Firft then, to fuppofe them to move in right Lines, you know is contrary to all the Laws and Principles we at prefent know of; and fince there are but two Ways that they can poffibly move in any natural Order, that is, either in right Lines, or in Curves, this being one, it muft of courfe be the other, *i. e.* in an Orbit; and confequently, were we able to view them from their middle Pofition, as from the Eye feated in the Center of *Plate* XXV. we might expect to find them feparately moving in all manner of Directions round a general Center, fuch as is there reprefented. It only now remains to fhew how a Number of Stars, fo difpofed in a circular Manner round any given Center, may folve the Phænomena before us. There are but two Ways poffible to be propofed by which it can be done, and one of which I think is highly probable; but which of the two will meet your Approbation, I fhall not venture to determine, only here inclofed I intend to fend you both. The firft is in the Manner I have above de-fcribed, *i. e.* all moving the fame Way, and not much deviating from the fame Plane, as the Planets in their heliocentric Motion do round the folar Body. In this Cafe the primary, fecondary, and tertiary conftituent Orbits, *&c.* framing the Hypothefes, are reprefented in *Plate* XXII, and the Confequence of fuch a Theory arifing from fuch an univerfal Law of Mo-tion

tion in *Plate* XXIII. where B, D denotes the local Motion of the Sun in the true *Orbis Magnus*, and E, C that of the Earth in her proper fecondary Orbit, which of courfe is fuppofed, as is fhewn in the Figure to change its fidereal Pofitions, in the fame Manner as the Moon does round the Earth, and confequently will occafion a kind of Proceffion, or annual Variation in the Place of the Sun, not unlike that of the Equinoxes, or Motion of all the Stars together, from Weft to Eaft round the Ecliptic Poles, and probably may in fome Degree be the Occafion of it. This Angle is reprefented, but much magnified, by the Lines F, C, G, and the Unnaturalnefs, or Abfurdity of a right Line Motion of the Sun by the Line I, H.

The fecond Method of folving this Phænomena, is by a fpherical Order of the Stars, all moving with different Direction round one common Center, as the Planets and Comets together do round the Sun, but in a kind of Shell, or concave Orb. The former is eafily conceived, from what has been already faid, and the latter is as eafy to be underftood, if you have any Idea of the Segment of a Globe, which the adjacent Figures, will, I hope, affift you to. The Doctrine of thefe Motions will perhaps be made very obvious to you, by infpecting the following Plates.

PLATE XXIV.

Is a Reprefentation of the Convexity, if I may call it fo, of the intire Creation, as a univerfal Coalition of all the Stars confphered round one general Center, and as all governed by one and the fame Law.

PLATE XXV.

Is a centeral Section of the fame, with the Eye of Providence feated in the Center, as in the virtual Agent of Creation.

PLATE XXVI.

Reprefents a Creation of a double Conftruction, where a fuperior Order of Bodies C, may be imagined to be circumfcribed by the former one A, as poffeffing a more eminent Seat, and nearer the fupream Prefence, and confequently of a more perfect Nature. Laftly,

PLATE XXVII.

Reprefents fuch a Section, and Segments of the fame, as I hope will give you a perfect Idea of what I mean by fuch a Theory.

Fig. 1. is a correfponding Section of the Part at A, in *Fig.* 2. whofe verfed Sine is equal to half the Thicknefs of the ftarry Vortice A C, or B A. Now I fay, by fuppofing the Thicknefs of this Shell, 1. you may imagine the middle Semi-Chord A D, or A E, to be nearly 6; and con-

PLATE XXII.

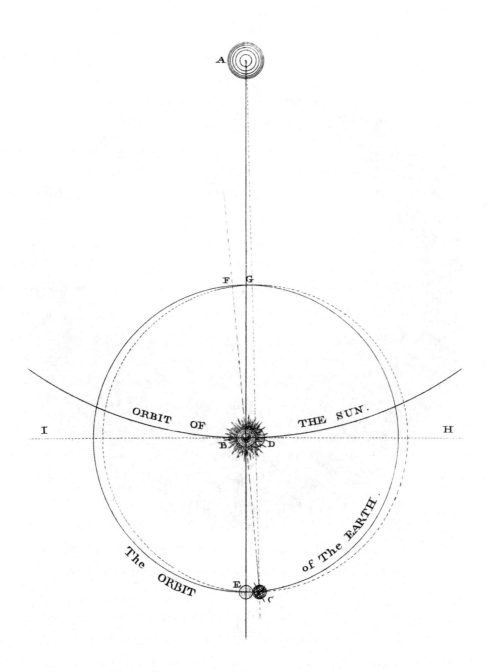

PLATE XXIV.

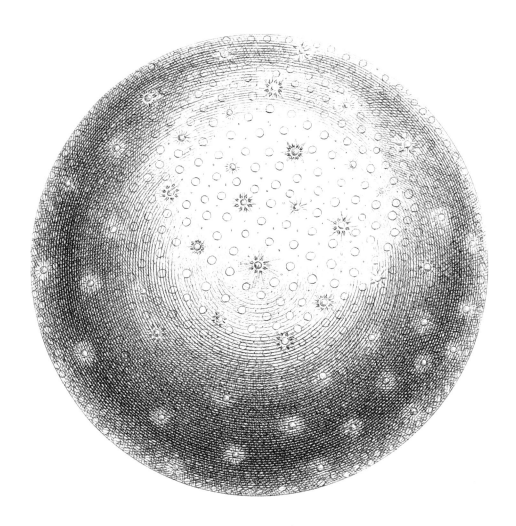

PLATE XXV.

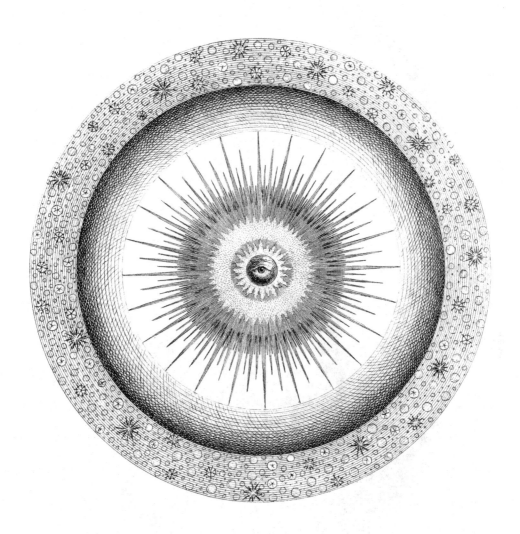

PLATE. XXVI.

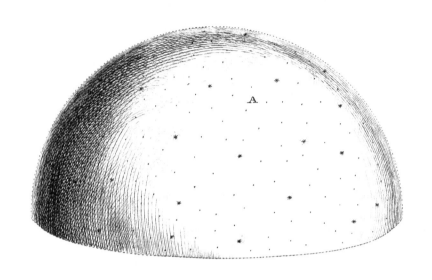

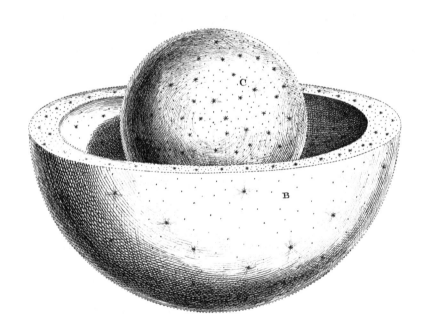

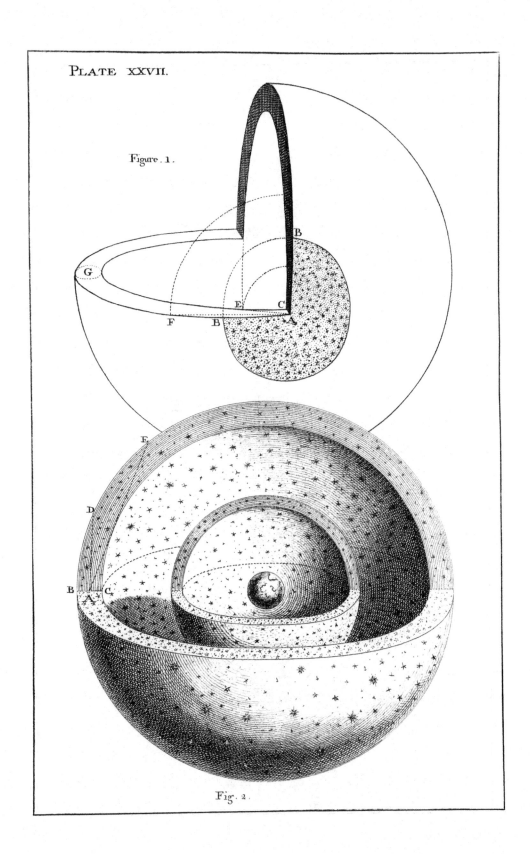

PLATE XXVII.

Figure. 1.

Fig. 2.

PLATE XXVIII.

Figure I.

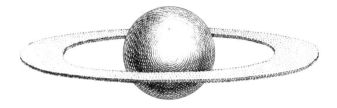

Fig. II.

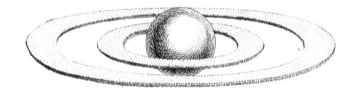

Fig. III.

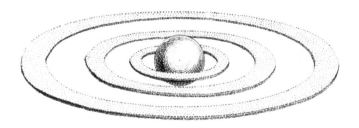

Figure I .

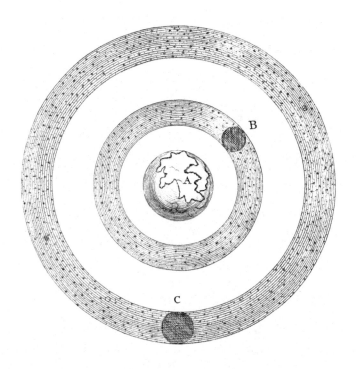

Fig. II .

thus in a like regular Diftribution of the Stars, there muft of courfe be at leaft three Times as many to be feen in this Direction of the Sine, or Semi-chord A E, itfelf, than in that of the femi verfed Sine A C, or where near the Direction of the Radius of the Space G. *Q. E. D.*

But we are not confined by this Theory to this Form only, there may be various Syftems of Stars, as well as of Planets, and differing probably as much in their Order and Diftribution as the Zones of *Jupiter* do from the Rings of *Saturn*, it is not at all neceffary, that every collective Body of Stars fhould move in the fame Direction, or after the fame Model of Motion, but may as reafonably be fuppofed as much to vary, as we find our Planets and Comets do.

Hence we may imagine fome Creations of Stars may move in the Direction of perfect Spheres, all varioufly inclined, direct and retrograde; others again, as the primary Planets do, in a general Zone or Zodiack, or more properly in the Manner of *Saturn*'s Rings, nay, perhaps Ring within Ring, to a third or fourth Order, as fhewn in *Plate* XXVIII. nothing being more evident, than that if all the Stars we fee moved in one vaft Ring, like thofe of *Saturn*, round any central Body, or Point, the general Phænomena of our Stars would be folved by it; fee *Plate* XXIX. *Fig.* 1. and 2. the one reprefenting a full Plane of thefe Motions, the other a Profile of them, and a vifible Creation at B and C, the central Body A, being fuppofed as *incognitum*, without the finite View; not only the Phænomena of the *Milky Way* may be thus accounted for, but alfo all the cloudy Spots, and irregular Diftribution of them; and I cannot help being of Opinion, that could we view *Saturn* thro' a Telefcope capable of it, we fhould find his Rings no other than an infinite Number of leffer Planets, inferior to thofe we call his Satellites: What inclines me to believe it, is this, this Ring, or Collection of fmall Bodies, appears to be fometimes very excentric, that is, more diftant from *Saturn*'s Body on one Side than on the other, and as vifibly leaving a larger Space between the Body and the Ring; which would hardly be the Cafe, if the Ring, or Rings, were connected, or folid, fince we have good Reafon to fuppofe, it would be equally attracted on all Sides by the Body of *Saturn*, and by that means preferve every where an equal Diftance from him; but if they are really little Planets, it is clearly demonftrable from our own in like Cafes, that there may be frequently more of them on one Side, than on the other, and but very rarely, if ever, an equal Diftribution of them all round the *Saturnian* Globe.

How much a Confirmation of this is to be wifhed, your own Curiofity may make you judge, and here I leave it for the Opticians to determine. I fhall content myfelf with obferving that Nature never leaves us without

K

a fuffi-

a fufficient Guide to conduct us through all the neceffary Paths of Knowledge; and it is far from abfurd to fuppofe Providence may have every
where throughout the whole Univerfe, interfperfed Modules of every
Creation, as our Divines tell us, Man is the Image of God himfelf.

Thus, Sir, you have had my full Opinion, without the leaft Referve,
concerning the vifible Creation, confidered as Part of the finite Univerfe;
how far I have fucceeded in my defigned Solution of the *Via Lactea*,
upon which the Theory of the Whole is formed, is a Thing will hardly
be known in the prefent Century, as in all Probability it may require fome
Ages of Obfervation to difcover the Truth of it.

It remains that I fhould now give you fome Idea of Time and Space;
but this will afford Matter fufficient for another Letter.

I am, &c.

LETTER

LETTER the EIGHTH.

Of Time and Space, with regard to the known Objects of Immensity and Duration.

SIR,

THE Opportunity you gave me in your laſt Viſit, of ſhewing you my general Scheme of the Univerſe, I find, beſides the Pleaſure it then gave, is now attended with many uſeful Advantages.

I now not only hope to be better underſtood for the future, but have reaſon to expect what I now write will merit your Attention more, and have ſome Title to your Approbation. The Ideas I have fram'd of Time and Space, will now more gradually fill your Imagination both with Wonder and Delight, before they can ariſe ſo high as to be loſt in an Eternity and the Infinity of Space. And I am fully perſwaded your farther Inquiries into theſe vaſt Properties of the Deity, will here be anſwered intirely to your Satisfaction. You muſt allow me now to be in ſome meaſure a Judge of what I think will pleaſe you moſt, from the Obſervations you have made upon my general Syſtem, or otherwiſe you would have reaſon to think me perhaps too preſuming : But I flatter myſelf the great Difficulty is now over ; and what remains to be ſaid, will alſo naturally follow from what has gone before, that this Letter, I gueſs, will go near to furniſh you with all the Ideas you wiſh to form upon the Subject. To what you have ſaid of my having left out my own Habitation in my Scheme of the Univerſe, having travell'd ſo far into Infinity as both to loſe ſight of, and forget the Earth, I think I may juſtly anſwer as *Ariſtotle* did when *Alexander*, looking over a Map of the World, enquir'd of him for the City of *Macedon* ; 'tis ſaid the Philoſopher told the Prince, That the Place he ſought for was much too ſmall to be there taken Notice of, and was not wit out ſufficient Reaſon omitted.

The Syſtem of the Sun compar'd but with a very minute Part of the viſible Creation, takes up ſo ſmall a Portion of the known Univerſe, that in a very finite View of the Immenſity of Space, I judg'd the Seat of the Earth to be of very little Conſequence, could I have poſſibly repreſented it, as not only being one of the ſmalleſt Objects in our Regions, but in a

manner

manner infinitely less than even her own annual Orbit, and had nothing to do with my main Defign, which was to reprefent all our planetary Worlds as one collective Body, and begin my comparative Scale of Magnitude from the Sun only and his Sphere of activity; as the fmalleft Object I could with any Propriety pretend to exprefs in fuch a Plan.

In fome Meafure to convince you that I have committed no Error in this, I will try by fome lefs mathematical Method than that of meer Numbers, to imprint an Idea in your Mind of the true Extent of the folar Syftem, and the Magnitude of all its moving Bodies, by natural Objects moft familiar to your Senfes. When we endeavour to form any Idea of Diftance, Magnitude, or Duration, by Numbers only, we fo foon exceed the Limits of Conception, that this way we find our Faculties of reafoning as finite as our Senfes; and no doubt 'tis right it fhould be fo, Providence, as it were, having ordain'd that the firft fhould only attend the laft, in fuch an adequate Degree to a determin'd Diftance; but what Diftance or Degree of Knowledge is deftin'd to human Nature, none but the Power that gave it can tell. 'Tis certain that beyond the third or fourth Place of our Nomenclator, we receive but very faint Impreffions of the thing expreft, and can frame fcarce any Notion at all of either Number, Diftance, or Magnitude, fignified beyond it : Hence Aftronomers are frequently oblig'd to have recourfe to mixt Ideas, and make Things of different Natures and Properties affift each other, to excite more adequate Ideas of what they would have conceived. Thus to exprefs immenfe Diftances and Magnitude, they frequently apply themfelves to Time and Motion; and *vice verfa*, to fignify a long Duration, they have often recourfe to Diftance and Matter, removing, in Imagination, Worlds of Sand, Grain after Grain, to fome remote known Region.

Hefiod, * to exprefs his Idea of the Diftance from his higheft Heaven to Earth, and from Earth to Hell, or *Tartarus*, fuppofes an Anvil to be let fall from one to the other, which he fays in nine natural Days would reach the Earth from Heaven, and in the fame time would fall from the Earth to Hell. † *Homer* makes his *Vulcan* fall from Heaven to the Ifland of *Lemnos* in much lefs Time, not exceeding one full artificial Day.

* From the high Heaven a brazen Anvil caft,
Nine Nights and Days in rapid Whirls would laft,
And reach the Earth the Tenth, whence ftrongly hurl'd;
The fame the Paffage to th' infernal World.

COOKE.

† Hurl'd headlong downward from th' etherial Height;
Tofs'd all the Day in rapid Circles round,
Nor till the Sun defcended touch'd the Ground.

POPE.
Modern

Modern Aſtronomers have made uſe of the ſwifteſt Velocity of a Cannon-Ball as continued thro' the Space they would ſo deſcribe, and in this Light, the Diſtance to the Sun has been by many compar'd to twenty-five Years Motion of a Cannon-Ball, ſuppoſing it to travel at the Rate of 100 Fathom in a Moment, *i. e. the Pulſe of an Artery*; and that a Journey ſo performed to one of the neareſt fix'd Stars, would take the ſame Body at leaſt 100,000 Years before it could arrive there. But the Method I have choſe to convey my Ideas of the Magnitude of the planetary Bodies, and the Extent of the viſible Creation to you, I am willing to hope you will find ſtill more familiar, comprehenſive, and eaſy: And it only depends upon your Remembrance of a very few known Objects, and their neighbouring Diſtances, which may be preſumed you are, or have been, very well acquainted with. You have not only very lately but very often been in *London*, and muſt, I think, retain ſome Idea of the Dome of St. *Paul*'s, tho' I own I ought not to be ſorry if you ſhould chance to have forgot it, provided it might prove a Means of making your Viſits more frequent. The Diameter of the Dome of this Church is 145 Feet: Now if you can imagine this to repreſent the Surface of the Sun, a ſpherical Body 18 Inches diameter, will juſtly repreſent the Earth in like Proportion; and another of only five Inches diameter, will repreſent the Moon. The Truths of theſe Proportions I have ſhewn in my *Clavis Cœleſtis*; and the Reaſon why I have here fixt upon the Dome of this Church for my firſt Object of Compariſon, will naturally appear from what follows.

From the Magnitude of the Earth on which we live, as from a known Scale with reſpect to its Parts compar'd with our own Bodies, we naturally frame our firſt Ideas of Extent, and fix our Rationale of Remoteneſs; by which we are ſufficiently enabled to judge of all other ſenſible Diſtances within one finite View. And hence by the undoubted Principles of Geometry, having firſt given the Meaſurement of the Earth in any known Proportion with any other Quantity moſt familiar to our Senſes, and the Angle of Appearance, or Parallax to any perceivable Object, we can eaſily find in homogenial Parts its true Diſtance from the Eye. And thus allowing for ſome ſmall tho' unavoidable Errors, that may poſſibly ariſe from the Difficulties of Obſervation (eſpecially ſmall Angles and minute Quantities) we can always determine to a ſufficient, and very frequently to a juſt Exactneſs, the relative Diſtance of all viſible Bodies, remote or near, ſuch as the Planets, Comets, and the Sun.

* In this Manner Aſtronomers having procur'd a comparative Standard, reduc'd to ſome known Meaſure, as *Engliſh* Miles, Leagues, Semi-Orbs or Orbits,

---

* Parallax is the changeable Poſition of Bodies to different Situations of the Eye. Firſt having found the Quantity of a Degree (*i. e.* a 60th Part of the Circumference) upon the Earth's

Orbits, with all the Force of analogical Reafoning, clearly can demonftrate the Place and Diftance of any Object within the Reach of Obfervation, and judge of Diftances almoft indefinite.

PLATE XXX.

Will help you to very correct Ideas of the real Magnitude of the Globe of the Earth, compar'd with the juft Extent of the Ifland of *Great-Britain*, which you will find with *Ireland*, and the reft of its Iflands, feated near the Center of the Projection. This as a Standard will enable you to judge of all other Diftances more perfectly; and firft I fhall confider that of the Sun.

The Sun is found to be mean diftant from the Earth nearly 81 Millions of Miles, or 6877,5 Diameters of the Earth; and *Saturn*, the remoteft Planet from him is at his greateft Diftance from us about 858 Millions of Miles: Yet thefe Diftances are but the beginning of Space, and only ferve to open our Ideas for farther Search.

The great Comet of 1680, as I have fome where faid before, was found to move in fo vaft an excentrick Orbit, that in its aphelion Point it would be 14,4 Times as far from the Sun, as the Orbit of *Saturn*, and hence at leaft eleven thoufand and two hundred Millions of Miles from us. Now fince the wife Creator hath fo difpos'd all the independent Parts of the Creation, fuch as the feveral Syftems of primary and fecondary Planets, &c. at fo great a Diftance from each other, that the Laws of any one in no wife fhall interfere, difturb, or interrupt the Principles of another; this Comet, which we can eafily prove belong'd to our own Sun, we may well imagine came not near any other; and tho' at that vaft Diftance from the folar Body, yet ftill there muft have remain'd a Space fufficient to divide or feperate the fenfible activity of neighbouring Syftems, that they may not rufh upon each other. Hence we may reafonably fuppofe, that the neareft Star can be no nearer than a triple Radius of its active Sphere; and provided they are all in regular Order, and much of the fame Magnitude with one another (which no Arguments can poffibly contradict) this Radius we may juftly make 2000 times the Diftance of our Earth. For admitting the utmoft Limits of the Sun's Attraction to exceed this Sphere of the Comets, as far as the Sphere of the Comets

Earth's Surface, *Aratofthenes* difcover'd that the Magnitude of the whole was eafily known; and then from the Moon's horizontal Parallax having given the Radius of the Earth, the Diftance of the Moon is foon determined: next by the menftrual Parallax of the Lunar Orbit, the Diftance of the Sun is found; and by the Elongation of the inferior Planets, their mutual Diftance from each other; and, laftly, from the annual Parallax of the Earth's Orbit, all the other Orbits of the fuperior Planets are eafily found.

exceeds

THE

EARTH

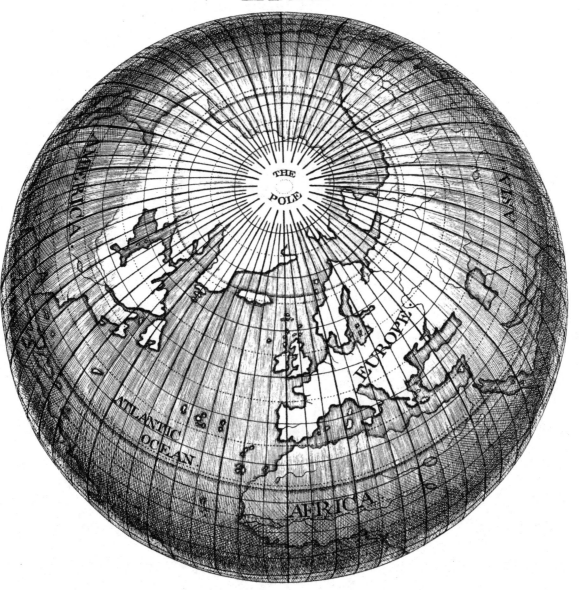

PLATE XXX.

exceeds that of the Planets, which is nearly 14,4 times, the Radius of the folar Syftem will be extended every way 200 Radius's of the Orbit of *Saturn*, and confequently the Diftance from Star to Star will not be lefs than 6000 times the Radius of our *Orbis Magnus*, and confequently upwards of 480,000,000,000 Miles. That this is even lefs than the real Truth, and may be defended as a very moderate Computation, grounded upon Reafon, we have infallible Demonftration to witnefs, and make appear as thus.

We know from the Nature of Diftance and Motion that the Stars may have an annual Parallax, but it is fo very fmall, that the very beft Aftronomers have never yet been able to affign what the Quantity really is. Yet it is allow'd by univerfal Confent, that it can't poffibly be more that one Minute of a Degree, and may probably be much lefs. Mr. *Flamftead*, by repeated Obfervations, made it in fome of them upwards of 40″; but Mr. *Bradley* has endeavour'd to prove it is every where too fmall to be determined, and affigns this Angle to another Caufe. This way then we cannot make their Diftance lefs; and to prove that it is fomething more than I have faid it is, let us even increafe the doubtful Parallax of 40″ to the moft it poffibly can be, *viz.* to 60″ or 1′; and by the Solution of the Triangle, we fhall find that the neareft Star is 6875 times the Radius of the Earth's Orbit from the Sun: And this tho' more than any other Proportion makes them, is ftill undeniably lefs than the Truth, which every Mathematician will of courfe be convinc'd of; and you yourfelf of force muft believe, when you are told, that the fmaller the Angle of Parallax is, the farther the Body is remov'd from us. By which Rule, according to Mr. *Flamftead's* Obfervations, the Diftance muft be ftill greater: By the optical Experiment of * Mr. *Huygins*, greater ftill than this; and according to Mr. *Bradley*, fo much more as not even too be determin'd.

Now if the reft are in general from each other, allowing the fame Extent of Syftem, and as much to part the like Extreams of active Virtue, be in fuch Proportion of aerial Space, it will appear, that to pafs from any one Star to another, we muft fly thro' fo vaft a Tract of pure Expanfe or Ether, that to vifit any one of the moft neighbouring Syftems, could we travel even as faft as the fwifteft Eagle flies, for Inftance, 500 Miles *per* Day, yet fhould we be 3,000,000 of Years upon our way before we could arrive there; and if continuing on to view the Regions of the reft within the known Creation, Myriads of Ages would be fpent, and yet we could not hope to fee the whole of but the fmalleft Conftellation.

* 27664 Radius's of the *Orbis Magnus*, equal to the Diftance of *Syrius*, whofe Parallax fhould be to anfwer it but 14″ 48‴.

But

But what Idea of Diſtance can you receive from this ſort of Eſtima-tion, where Numbers ariſe ſo very high. I own to you mine are ſoon quite loſt by this Method of counting, either, Diſtances or Duration. I believe few People can range their Ideas with ſuch Perſpicuity, as to arrive at any adequate Notion of any Number above a thouſand.

To give you therefore a clearer Idea of Diſtance, and impreſs the Pro-portions of Space more ſtrongly and fully in your Mind, let us ſuppoſe the Body of the Sun, as I have ſaid before, to be repreſented by the Dome of St. *Paul's*; in ſuch Proportion a ſpherical Body eighteen Inches Dia-meter, moving at *Mary-le-bone*, will juſtly repreſent the Earth, and ano-ther of five Inches Diameter, deſcribing a Circle of forty-five Feet and a half Radius round it, will repreſent the Orbit and Globe of the Moon. A Body at the *Tower* of 9,7 Inches, will repreſent *Mercury*; and one of 17,9 Inches at St. *James's* Palace will repreſent the Planet *Venus*; *Mars* may be ſuppoſed at a Diſtance, like that of *Kenſington* or *Greenwich*, 10 Inches Diameter: *Jupiter*, imagined to be at *Hampton-Court*, or *Dartford* in *Kent*; and *Saturn*, at *Clieſden*, or near *Chelmsford*: The firſt repre-ſented by a Globe 15 Foot 4 Inches Diameter, the latter by one of 11 Feet ½ and his Ring four Feet broad: Theſe would all naturally repreſent the planetary Bodies of our Syſtem in their proper Orbits and proportional Magnitudes, as moving round the Cupola of St. *Paul's*, as their common Center the Sun. And preſerving the ſame natural Scale, the Aphelion of the firſt Comet would be about *Bury*, the ſecond at *Briſtol*, and the third near the City of *Edinburgh*. But if you will take into your Idea one of the neareſt Stars; inſtead of the Dome of St. *Paul's*, you muſt ſuppoſe the Sun to be repreſented by the gilt Ball upon the Top of it, and then will another ſuch upon the Top of St. *Peter's* at *Rome* repreſent one of the neareſt Stars.

The whole Syſtem exhibited in the above Proportion, would be nearly as follows :

Diameter of the Sun   145 Feet.
           Saturn   11,587, his Ring 27,54, its Breadth 4.
           Jupiter, 15,39.
           Mars,   10,15 Inches.
       the Earth,   18,125.
           Venus,   17,98
           Mercury, 9,715
   and the Moon,   4,93

Diſtance

* Diſtance of *Saturn* from the Sun, 27 Miles, and 1700 Yards.

            Jupiter,   15 Miles, and 458 Yards.

            Mars,   4 Miles, and 751 Yards.

            the Earth,   2 Miles, and 1632 Yards.

            Venus,   2 Miles, and 217 Yards.

            Mercury,   1 Mile, and 267 Yards.

and of the Moon, from us, 45 Yards and a half.

That of the moſt diſtant Comet 390, and the neareſt of the Stars not leſs than 6875, † Radius's of the *Orbis Magnus*.

Now, if like Creations crowd the vaſt Depths of Infinity, and if each are adapted to receive Beings of different Natures, where muſt our Wonder and Ideas have end?

As it is evident in the Sign *Taurus,* in *Perſeus,* and *Orion,* that we can plainly perceive Stars to the ſixth and ninth Magnitude, the former with our naked Eye, the other by the Help of Teleſcopes, the viſional ocular Creation cannot be leſs than 4,320,000,000,000 Miles in ſemi Diameter, and admitting a regular Diſtribution of thoſe primordial Bodies amongſt themſelves the Depth, or moſt remote Limits of the *Vortex Magnus* from Side to Side, cannot be leſs than 8 m, m, 640 thouſand of Million of Miles, admitting it is no more than what we ſee; and laſtly, ſuppoſing our Syſtem to be ſituated nearly in the Middle of the *Vortex Magnus* (which, from the viſible Order of the Stars, we may juſtly conjecture, with the higheſt Probability of Truth) the neareſt Diſtance of the *Ens Primum,* in the Realms of eternal Day, will riſe to 30,000,000,000 000 Miles, but more probably to 100,000,000,000,000 Miles, making the Confines of Creation from Verge to Verge in the firſt Caſe, upwards of 68 Million of Millions of Miles, Diameter, and by the laſt above 200. But, if we compute the Diſtance of the Stars after the Manner of *Huygens,* for his Diſtance of *Syrius* from the Sun, the Diſtance of the Region of Immortality without exceeding Probability may riſe to near 1,000,000,000,000,000 Miles.

Now to paſs by any progreſſive Motion from the outward Verge, or Borders of the Creation, thro' the ſtarry Regions of Mortality, if I may call

---

* Of the Satellites of *Saturn* in the above Proportion.   And thoſe of *Jupiter.*

The 1 would be 27,96

     2   35,52 } Feet diſtant from his Center.

     3   50,

     4   114,

     5   341,9

The 1 would be 28,51

     2   69,177 } Feet diſtant from him.

     3   110,224

     4   190,

* Radius, or Sign of   89   59   30 —— —— 10,0000000

Sine ſubſtract of   0   0   30 —— —— 6,1626961

Hence the Diſtance   6875,5 —— —— 3,8373039

them

them fo, as far as the Center of the *Ens Primum*, or *Sedes Beatorum*, according to *Homer*, or *Milton*'s Manner of meafuring Space, a Body falling, or a Being moving with a Velocity but of 1000 Feet *per* Minute, *i. e.* at the Rate of 20,000 Yards *per* Hour, or about 300 Miles *per* Day, would be at leaft 300,000,000 Years upon its Journey thither, if not 1,000, m, and perhaps much more, without offending Probability ; but even three Million Centuries, or Ages, fure is enough to be employ'd, in paffing from one Place to another ; therefore, we may conclude, the Soul muft have fome other Vehicle than can be found in the Ideas of Matter to convey it fo far, at leaft at once. Hence we may truly infer, that the Soul muft be immaterial, and that in all Probability there may be States in the Univerfe fo much more longer lived than ours, that, compared with the Age of Man, the Age of fuch Beings may be almoft as an Eternity, or rather, as that of the human Species to that of a Sun-born Infect.

Again, if there are ftill Stars beyond all thefe of other Denomination, which we do not here perceive, how vaftly muft thefe Numbers be increafed, to exprefs, almoft without Idea, the amazing Whole of this one vifible Creation ; but what has been already faid, I judge will be fufficient to fhow the Immenfity of Space, and help you to conceive the ftupendious Nature of an endlefs Univerfe ; every where the home Poffeffion, Production, and inftantaneous Care, of an infinite good Being, perfectly wife, and powerful, of whom we can have no Idea more, than a Being in dark Privation can have of Light, but through the Luftre of his own refplendent Attributes.

Thus, having attempted to enlarge your Ideas of the Creation in general, and in fome meafure having confidered the Indefinity of Space, I fhall in the next Place proceed to give you fome Account of my Notions of Time.

As Diftance is the Meafure of Magnitude and of all Extent, and helps our Imagination to the Ideas of Space, fo are progreffive Moments the Meafure of Velocity, and makes us fenfible of Duration : And as Space may be extended through all Infinity, fo Time may be continued as to Eternity. This Succeffion of temporal Ideas impreffed, or excited in the Mind, as an Effect of Matter in Motion, producing a perpetual Change, both of Objects earthly and celeftial, enables us not only to reflect upon paft Viciffitudes of Nature, but from their regular Courfes, known Order and Returns, predict Phænomena to come, and prove the periodical Effects of Nature's conftant Laws fo juft and certain, that Time may be faid with Truth, to co-exift with Motion.

Meafure being a certain Quantity of Senfation interwove with our Ideas of Diftance and Duration, proceeding from a Reflection of what is impreffed upon the Mind by fome external Object, I muft again return to our Mother of Ideas the Earth, and from thence, as I did, of Diftance,

frame

frame the original Images beft fuited to the Underftanding, proper for our Judgment of Duration.

Time takes its firft Denomination from the diurnal Rotation of the Earth upon its Axis, which we call a natural Day, and this for obvious Reafons we fubdivide in twenty-four Parts or Hours. This diurnal Motion having been fucceffively repeated, and the Day renewed three hundred and fixty-five Times, we find that all the vegetable World has gone through all its Variegations, and Nature has again put on the fame Face, adapted to the Seafon ; during which Time, and indeed which occafions this general Change and Repetition, the Earth is found to make one intire Revolution round the Sun. This Space, or Period of Time, we call a folar, or rather a natural Year ; and from our Senfibility of this, and its conftituent Parts, both horary and diurnal, we form our general Judgment of Duration.

Saturn, the moft remote, and moft regular Planet in our Syftem, as has been faid before, performs one Revolution round the Sun in about twenty-nine of the above folar Years: The great Comet of 1680 makes but one periodical Return in five hundred and feventy-five of thofe Years, and the general Motion of the Stars, arifing from the Proceffion of the Equinoxes, altogether continually changing their Afpect, or Pofition, at the Rate of 50 *per* Year round the ecliptic Poles, compleats but one Revolution in 25920 Years ; in which Time the whole fidereal Frame of Heaven has changed, and every Star returned to the fame Point of the folar Sphere it fet out from. This is by many called the great *Saturnian* Year : Concerning which, Mr. *Addifon* has thus tranflated an eminent Author.

> When round the great *Saturnian* Year has turn'd,
> In their old Ranks the wandering Stars fhall ftand,
> As when firft marfhall'd by the Almighty's Hand.
>
> ADDISON.

Now, if this fidereal Revolution, arifing from a fecondary Caufe, require this Number of Years to perfect one Rotation, what muft their primitive Orbits take to circumfcribe the *Vortex Magnus*.

It has been obferved, that the biggeft Star to us fcarce moves a Minute in an hundred Years, and the moft remote as infenfibly for Ages, from whence and what has been already faid of the imagined Diftance of the general Center, we may frame this probable and well-grounded Guefs, that the mean Revolution of a Star near the Middle of the *Vortex Magnus*, cannot be made in lefs than a Million of Years, and though to us imperceptible, our Sun in his own orbicular Direction, may be moving many Miles *per* Day. Befides, if local Motion can be proved amongft the Stars, what lefs than an Eternity can again reftore them to their original Order and primitive State.

L 2 Such

Such vaſt Room in Nature, as *Milton* finely expreſſes it, cannot be without its Uſe ; and nothing but abſolute Demonſtration is wanting (which from their Nature and Diſtance cannot be expected) to confirm the grand Deſign, ſo ſuited to the Deity's infinite Capacity, and of eternal Benefit to all his Creatures, eſpecially Beings of a rational Senſe, and in particular Mankind.

Of theſe habitable Worlds, ſuch as the Earth, all which we may ſuppoſe to be alſo of a terreſtrial or terraqueous Nature, and filled with Beings of the human Species, ſubject to Mortality, it may not be amiſs in this Place to compute how many may be conceived within our finite View every clear Star-light Night. It has already been made appear, that there cannot poſſibly be leſs than 10,000.000 Suns, or Stars, within the Radius of the viſible Creation ; and admitting them all to have each but an equal Number of primary Planets moving round them, it follows that there muſt be within the whole celeſtial Area 60,000,000 planetary Worlds like ours. And if to theſe we add thoſe of the ſecondary Claſs, ſuch as the Moon, which we may naturally ſuppoſe to attend particular primary ones, and every Syſtem more or leſs of them as well as here ; ſuch Satellites may amount in the Whole perhaps to 100,000,000, or more, in all together then we may ſafely reckon 170,000,000, and yet be much within Compaſs, excluſive of the Comets which I judge to be by far the moſt numerous Part of the Creation.

In this great Celeſtial Creation, the Cataſtrophy of a World, ſuch as ours, or even the total Diſſolution of a Syſtem of Worlds, may poſſibly be no more to the great Author of Nature, than the moſt common Accident in Life with us, and in all Probability ſuch final and general Doom-Days may be as frequent there, as even Birth-Days, or Mortality with us upon the Earth.

This Idea has ſomething ſo chearful in it, that I own I can never look upon the Stars without wondering why the whole World does not become Aſtronomers ; and that Men endowed with Senſe and Reaſon, ſhould neglect a Science they are naturally ſo much intereſted in, and ſo capable of inlarging the Underſtanding, as next to a Demonſtration, muſt convince them of their Immortality, and reconcile them to all thoſe little Difficulties incident to human Nature, without the leaſt Anxiety.

Such a Protheſis can ſcarce be called leſs than an ocular Revelation, not only ſhewing us how reaſonable it is to expect a future Life, but as it were, pointing out to us the Buſineſs of an Eternity, and what we may with the greateſt Confidence expect from the eternal Providence, dignifying our Natures with ſomething analogous to the Knowledge we attribute to Angels ; from whence we ought to deſpiſe all the Viciſſitudes of adverſe Fortune, which make ſo many narrow-minded Mortals miſerable.

I am now, &c.

LETTER the NINTH.

Reflections, by Way of General Scolia, *of Consequences relating to the Immortality of the Soul, and concerning Infinity and Eternity.*

SIR,

THIS my laft Letter to you, I mean my final aftronomical tone, I propofe as a *General Scolia* to the reft, the principle Matter being Reflections upon what is gone before, with fome Conclufion naturally following or appendant to what has been already faid; but which, I could not in any other Place, fo properly remark to you.

The Probability of the foregoing Conjectures, chiefly built upon very diftant Obfervations, fhew an apparent Neceffity for fome other kind of Doctrine permitted by Providence, to give Mankind a Knowledge of their Immortality and Dependance upon it, in the firft Ages of the World.

And for the fame Reafon it evidently appears, that the ancient Philofophers had it not in their Power to prove a fupream *Being* and Director of all Things this Way.

And yet, as by a Sort of Inftinct, or natural Reafon, and Confcioufnefs of a *good Principle*, we fee how many noble Steps they made towards it, and was convinc'd at laft of this *great Truth*, that fince there was a *Mind* in fo imperfect a Creature as Man, the *perfect Univerfe*, which comprehended all Things, could not poffibly be without one ; and as Sir *Ifaac Newton* has juftly obferved in his *Principia*, " If every Par-" ticle of Space be *always*, and every individual Moment of Duration " *every where*; furely the Maker and Lord of all Things, cannot be *never* " and *no where*."

To make manifeft the infinite Empire and Agency of God, from celeftial Motion, became the Tafk, but of very late Years ; and I can't help being of Opinion, that by means of thefe primary Bodies, only, we fhall at length be able to trace the greater Circulations, and Laws of Nature, to their real original and fountain Head.

Thefe

These, were any thing wanting, besides the *Miracle ourselves*, to convince us of a divine Origination, are all infallible Proofs, that the Universe is governed by an intelligent and all-powerful Being, whose Existence is too nearly related to a self-evident Truth to be more clearly demonstated, than it is manifest of itself, both from the particular Laws of Nature, and the general Order of Things. An Argument which has been thought of no small Force, and well worth observing in the Infancy of *Christianity*. *The invisible Things of God are clearly seen, being understood by the Things that are made, even his eternal Power and Godhead.* Rom. i. 20.

But 'tis now high time to look back upon my Theory, and tell you it is a vain Supposition, to imagine I shall ever be able to convince every Reader, either of the Truth or Probability of what I have advanced to you : Mathematical Assistance not being to be expected, where perhaps it has never been thought of ; and I allow you, it is much more reasonable to expect, that fifty Persons will read these Letters without perceiving the Reasonableness of them, than that five should consider them with proper Judgment.

I must ingenuously confess to you, that nothing is wanting to convince me intirely of the Certainty of what I here advance by way of Conjecture to you. But this you must only look upon as an happy Partiality, which generally attends all Authors, and always will be the chief Support of their tedious Labours. I assure you, I have neither Hopes nor Expectation, no, not the weak Breath of a Wish, to be admitted a proper Judge of my own Works. But I shall always take their Imperfection to be rather, (like my own Faults) to be too near me to be seen ; I therefore trust all to my Friend, and if I am so fortunate as to excite his Approbation, I shall think myself very happy in a very favourite Point ; which is, The advancing nothing which a rational Reader would willingly overlook or be ignorant of.

But if I have been so happy as to come so near the Mark, as to border upon Truth, I believe you will allow me to carry my Conjectures a little further, and point out some farther pleasing Consequences, which I begin to perceive may naturally follow.

Should it be granted, that the Creation may be circular or orbicular, I would next suppose, in the general Center of the whole an intelligent Principle, from whence proceeds that mystick and paternal Power, productive of all Life, Light, and the Infinity of Things.

Here the to-all extending Eye of Providence, within the Sphere of its Activity, and as omnipresently presiding, seated in the Center of Infinity, I would imagine views all the Objects of his Power at once, and every Thing immediately direct, dispensing instantaneously its enlivening Influence,

to the remoteft Regions every where all round. I know you'll fay Aftronomers are never to be fatisfied, and I muft own where there is fo much
rational Entertainment for the human Mind, and fo fuitable to the
true Dignity of God, and moft worthy of Man, it is not eafy to
know where to ftop in fuch a Scene of Wonders.

Having, I fay, once granted that all the Stars may move round one
common Center, I think it is very natural to one, who loves to purfue
Nature as far as we may, to enquire what moft likely may be in that
Center; for fince we muft allow it to be far fuperior to any other Point
of Situation in the known Univerfe, it is highly probable, there may be
fome one Body of fiderial or earthy Subftance feated there, where the divine
Prefence, or fome corporeal Agent, full of all Virtues and Perfections,
more immediately prefides over his own Creation. And here this primary Agent of the omnipotent and eternal Being, may fit enthroned, as
in the *Primum Mobile* of Nature, acting in Concert with the eternal
Will. To this common Center of Gravitation, which may be fuppofed
to attract all Vertues, and repel all Vice, all Beings as to Perfection
may tend; and from hence all Bodies firft derive their Spring of Action,
and are directed in their various Motions.

Thus in the *Focus*, or Center of Creation, I would willingly introduce
a primitive Fountain, perpetually overflowing with divine Grace, from
whence all the Laws of Nature have their Origin, and this I think would
reduce the whole Univerfe into regular Order and juft Harmony, and
at the fame time, inlarge our Ideas of the divine Indulgence, open our
Profpect into Nature's fair Vineyard, the vaft Field of all our future
Inheritance.

But what this central Body really is, I fhall not here prefume to fay,
yet I can't help obferving it muft of Neceffity, if the Creation is real and
not merely Ideal, be either a Globe of Fire fuperior to the Sun, or
otherwife a vaft terraqueous or terreftial Sphere, furrounded with an
Æther like our Earth, but more refined, tranfparent and ferene. Which
of thefe is moft probable, I fhall leave undetermined, and muft acknow-
ledge at the fame time, my Notions here are fo imperfect, I hardly dare
conjecture. 'Tis true, I have ventur'd to think it may be one of thefe,
and fince fo glorious a Situation can hardly be fuppofed without its proper Inhabitants, 'tis moft natural to conclude it may be the latter. In
the firft Cafe, befides our having no Idea of Beings exifting in Fire, it
would not, notwithftanding its Diftance, be fo eafy to account for its being
invifible; and fince the Luftre of the Stars are all innate, they could
receive no Benefit from it, and confequently fuch a Nature as a folar Compofition, muft in this Place be render'd ufelefs; but in the latter Suppofition

pofition of its being a dark Body, we have no Difficulty attending us, having feveral Inftances of like Bodies, moving round an opaque one. Now when we confider, that all thofe radient Globes, which adorn the Skies, thofe bright ætherial Sparks of elemental Fire, thick ftrewed like Seeds of Light all round our Hemifphere, are each to us the Embrio of a glorious Sun ; how awful and ftupendious muft that Region be, where all their Beams unite and make one inconceivable eternal Day ?

Though the Deity, fays a learned Writer " be effentially prefent thro' " all the Immenfity of Space, there is one Part of it in which he difco- " vers himfelf in a moft tranfcendent and vifible Glory. This is that Place " which is mark'd out in Scripture, under the different Appellations " of PARADICE ; *the third Heaven* ; *the Throne of* GOD, *and the Habi-* " *tation of his Glory.*"

THIS continues the fame Author, is " that Prefence of God, which " fome of the Divines call his glorious, and others his majeftick Pre- " fence."

It is here, and here only, as in the Center of his infinite Creations, where he refides in a fenfible Magnificence, and in the midft of thofe Splendors, which can Effect the Imagination of his Creatures ; and though the moft facred and fupreme Divinity be allowed as effentially prefent in all other Places as well as in this, as being a BEING whofe Center is every where, and Circumference no where; yet it is here only, or in fuch Senforium of his Unity, where he manifefts his corporeal Agency, as in the Foci of his infinite Empire over all created Beings. It is to this majeftick Prefence of GOD, we may apply thofe beautiful Expreffions of Scripture, " *Behold even to the Moon and it fhineth not* ; *yea the Stars* " *are not pure in his Sight.*"

" The Light of the Sun, and all the Glories of the World, on which " we live, are but as weak and fickly Glimmerings, or rather Dark- " nefs it felf, in Comparifon of thofe Splendors, which encompafs this " Throne of GOD."

> Here Heav'ns wide Realms an endlefs Scene difplays,
> And Floods of Glory thro' its Portals blaze ;
> The Sun himfelf loft in fuperior Light,
> No more renews the Day, or drives away the Night :
> The Moon, the Stars, and Planets difappear,
> And Nature fix't makes one eternal Year.

Here and here alone center'd in the Realms of inexpreffible Glory, we juftly may imagine that primogenial Globe or Sphere of all Perfections,

subject

subject to the Extreams of neither Cold nor Heat, of eternal Temperance and Duration.   Here we may not irrationally suppose the Vertues of the meritorious are at last rewarded and received into the full Possession of every Happiness, and to perfect Joy.   The final and immortal State ordain'd for such human Beings, as have passed this Vortex, of Probation thro' all the Degrees of human Nature with the supream Applause.

What vast room is here, for infinite Power and Wisdom to act in, and that so visibly takes Delight to bless all his Beings with his Bounty.   And endless as his Prescience, Attributes, and Goodness, are undoubtedly all those natural and apparent Joys with which he manifests his Love to all his Creatures, a Multiplicity of Objects not to be enumerated.   For wheresoever we turn our Eyes, and follow with our Reason, we may meet with Worlds of all Formations, suited no doubt to all Natures, Tastes, and Tempers, and every Class of Beings.

Here a Groupe of Worlds, all Vallies, Lakes, and Rivers, adorn'd with Mountains, Woods, and Lawns, Cascades and natural Fountains; there Worlds all fertile Islands, cover'd with Woods, perhaps upon a common Sea, and fill'd with Grottoes and romantick Caves.   This Way, Worlds all Earth, with vast extensive Lawns and Vistoes, bounded with perpetual Greens, and intersperfed with Groves and Wildernesses, full of all Varieties of Fruits and Flowers.   That World subsisting perhaps by soft Rains, this by daily Dews, and Vapours; and a third by a central, subtle Moisture arising like an Effluvia, through the Pores and Veins of the Earth and exhaling or absorbing as the Season varies to answer Nature's Calls. Round some perhaps, so dense an Atmosphere, that the Inhabitants may fly from Place to Place, or be drawn through the Air in winged Chariots, and even sleep upon the Waves with Safety; round others possibly, so thin a fluid, that the Arts of Navigation may be totally unknown to it, and look'd upon as impracticable and absurd, as Chariot flying may be here with us; and some where not improbably, superior Beings to the human, may reside, and Man may be of a very inferior Class; the second, third, or fourth perhaps, and scarce allow'd to be a rational Creature. Worlds, with various Moons we know of already; Worlds, with Stars and Comets only, we equally can prove is very probable; and that there may be Worlds with various Suns, is not impossible.   And hence it is obvious, that there may not be a Scene of Joy, which Poetry can paint, or Religion promise; but somewhere in the Universe it is prepared for the meritorious Part of Mankind.   Thus all Infinity is full of States of Bliss; Angelic Choirs, Regions of Heroes, and Realms of Demi-Gods; Elysian Fields, Pindaric Shades, and Myriads of inchanting Mansions,

not

not to be conceived either by Philofophy or Fancy, affifted by the ftrongeft Genius and warmeft Imagination.

All harmonioufly crowded and provided with every Object of Beatitude, that Friendfhip, Love, or Society can infpire, the Mufes or the Graces Frame; and all as permanent and perfect, that is deftin'd to a Duration, fuited to the Nature of their Exiftence and Degree of Cognifance; for as a very learned Writer obferves upon this fame Subject;

" How can we tell, but that there may be above us Beings of greater
" Powers, and more perfect Intellects, and capable of mighty Things,
" which yet may have corporeal Vehicles as we have, but *finer* and
" *invifible?* Nay, who knows, but that there may be even of thefe
" many *Orders*, rifing in Dignity of Nature, and Amplitude of Power,
" one above another? It is no Way below the Philofophy of thefe Times,
" which feems to delight in inlarging the Capacities of Matter, to affert
" the Poffibility of this."

From thefe amazing Ideas of Space in general, and from the particular Diftance of the Stars, which feparates as it were, one Syftem of Bodies from another, and by fo prodigious an extent, as fcarce to be fuppos'd a temporal Tafk. I think it naturally follows, had we no other Way to prove it, or any other Reafon to believe it, that the Soul muft of Neceffity be immaterial; for as this Space feems fo impaffible to Matter, as not to be undertaken and performed without the Lofs of Ages, in a State only of Tranfmigration, we may well imagine, that Change of Place is not effected this Way, but by fome other Vertue or Property, more immediate, if not inftantaneous.

I own next to *Annihilation* is the State of Oblivion, and this Way we may folve all Difficulties with regard to our being fenfible of fuch a Lofs of Exiftence; but if we allow the Soul to be immaterial, it no longer has any thing to do with Space, but as operating by Reflection only, or the Faculty of Thinking; it may be like the Imagination where it pleafes in a Moment.

Objects of the Mind abftracted from the Senfes of the Body, has no real or comparative Magnitude; that is, I would fay, an Inch, a Foot, a Yard, a Mile, or a Million of Miles are all equally indefinite, and is thus prov'd; every finite Line is formed of an infinite Number of Points, and no finite Line can be folv'd into more. Thus if you will allow me the Expreffion, the Mind being magnified as all Objects are diminifhed, what feems impracticable in the natural State of Things, in an Ideal one, becomes very poffible; that is, to make myfelf more intelligible, though we can hardly conceive, how any Being can pafs from *Syrius* to the Sun, by natural Laws in their proper State, yet if proportionally reduced by a

new

PLATE. XXXI.

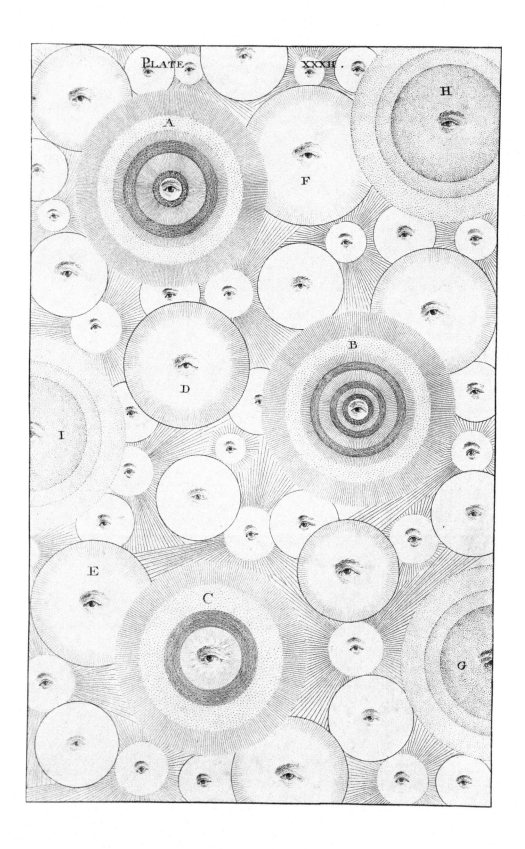

new Modification of Ideas, to the Bignefs of a Ball 6 Feet Diameter, and to be only 680 Miles afunder; the Thing is very comprehenfive and eafy.

Hence Vifion, Light, and Electrical Virtue, feem to be propagated with fuch Velocity, that nothing but God can poffible be the Vehicle; and hence we may juftly fay with St. *Paul, Acts* xvii. 28. *In him we live, in him we move, in him we have our Being.*

It will further appear, from the foregoing Letters, that all the Stars and planetary Bodies within the finite View, are altogether but a very minute Part of the whole rational Creation; I mean that vaft collective Body of habitable Beings, which I have endeavoured to demonftrate, are all govern'd by the fame Laws, though varioufly revolving round one common Center, in which Center we may not impertinently venture to fuppofe the prime Agent of our Natures; or otherwife, the moft perfect of all created Beings, illimitable in his Ideas and Faculties of Senfation particularly prefide.

> But tho' paft all diffus'd, without a Shore
> His Effence; *local* is his Throne, (as meet)
> To gather the difperft, (as Standards call
> The lifted from afar) to fix a Point;
> A central Point, collective of his Suns,
> Since finite ev'ry Nature, but his own.　　　Dr. *Young.*

And farther fince without any Impiety; fince as the Creation is, fo is the Creator alfo magnified, we may conclude in Confequence of an Infinity, and an infinite all-active Power; that as the vifible Creation is fuppofed to be full of fiderial Syftems and planetary Worlds, fo on, in like fimilar Manner, the endlefs Immenfity is an unlimited Plenum of Creations not unlike the known Univerfe. See *Plate* XXXI. which you may if you pleafe, call a partial View of Immenfity, or without much Impropriety perhaps, a finite View of Infinity, and all thefe together, probably diverfified; as at A, B and C. in *Plate* XXXII. which reprefents their Sections, if all may be a proper Term for an infinite or indefinite Number, we may juftly imagine to be the Object of that incomprehenfible Being, which alone and in himfelf comprehends and conftitutes fupreme Perfection.

That this in all Probability may be the real Cafe, is in fome Degree made evident by the many cloudy Spots, juft perceivable by us, as far without our ftarry Regions, in which tho' vifibly luminous Spaces, no one Star or particular conftituent Body can poffibly be diftinguifhed; thofe in

all

all likelyhood may be external Creation, bordering upon the known one, too remote for even our Telescopes to reach.

With the raptur'd Poet may we not justly say

> O, what a Root! O what a Branch is here!
> O what a Father! what a Family!
> Worlds! Systems! and Creations!

And in Consequence of this

> In an Eternity, what Scenes shall strike?
> Adventures thicken? Novelties surprize?
> What Webs of Wonder shall unravel there?
>
> *Night Thoughts.*

So many varied Seats where every Element may have its proper Beings and all adapted to partake of every thing suited to their Natures, argue such Maturity of Wisdom, and the vast Production such mysterious Power; 'tis hardly possible for Mortals not to see divine Intelligence preside, and that every Being somewhere must be happy

A Universe so well designed, and fill'd with such an endless Structure of material Beings, and all the Result of Prescience and infinite reflected Reason, flowing from a Mind all perfect, full of all Ideas, could never be designed in vain; and tho our narrow Bounds of Reason limited, by finite Senses, cannot directly see the Consequence dependant on a Sequel, yet from what we do see, great Room we have to hope the next Stage of Existence will be more lasting and more perfect; and it is highly probable, the noblest Suggestion of the most luxuriant Fancy may fall infinitely short of what we are designed for.

But here, even in this World, are Joys which our Ideas of Heaven can scarce exceed, and if Imperfection appear thus lovely, what must Perfection be, and what may we not expect and hope for, by a meritorious Acquiescence in Providence, under the Direction, Indulgence, and Protection of infinite Wisdom and Goodness, who manifestly designs perfect Felicity, as the Reward of Virtue in all his Creatures, and will at proper Periods answer all our Wishes in some predestined World.

All this the vast apparent Provision in the starry Mansions, seem to promise: What ought we then not to do, to preserve our natural Birthright to it and to merit such Inheritance, which alas we think created all to gratify alone, a Race of vain-glorious gigantick Beings, while they are confined to this World, chained like so many Atoms to a Grain of Sand.

I am, &c.

THE END.

Printed in the United States
By Bookmasters